A Brief History of the Universe

(and our place in it)

Sarah Alam Malik

**SIMON &
SCHUSTER**

London · New York · Amsterdam/Antwerp · Sydney/Melbourne · Toronto · New Delhi

First published in Great Britain by Simon & Schuster UK Ltd, 2026
Copyright © Sarah Alam Malik, 2026

The right of Sarah Alam Malik to be identified as the author of this work has been asserted in accordance with the Copyright, Designs and Patents Act, 1988.

1 3 5 7 9 10 8 6 4 2

Simon & Schuster UK Ltd
1st Floor
222 Gray's Inn Road
London WC1X 8HB

For more than 100 years, Simon & Schuster has championed authors and the stories they create. By respecting the copyright of an author's intellectual property, you enable Simon & Schuster and the author to continue publishing exceptional books for years to come. We thank you for supporting the author's copyright by purchasing an authorized edition of this book.

No amount of this book may be reproduced or stored in any format, nor may it be uploaded to any website, database, language-learning model, or other repository, retrieval, or artificial intelligence system without express permission. All rights reserved. Enquiries may be directed to Simon & Schuster, 222 Gray's Inn Road, London WC1X 8HB or RightsMailbox@simonandschuster.co.uk

www.simonandschuster.co.uk
www.simonandschuster.com.au
www.simonandschuster.co.in

Simon & Schuster Australia, Sydney
Simon & Schuster India, New Delhi

The authorised representative in the EEA is Simon & Schuster Netherlands BV, Herculesplein 96, 3584 AA Utrecht, Netherlands. info@simonandschuster.nl

The author and publishers have made all reasonable efforts to contact copyright-holders for permission, and apologise for any omissions or errors in the form of credits given. Corrections may be made to future printings.

Simon & Schuster strongly believes in freedom of expression and stands against censorship in all its forms. For more information, visit BooksBelong.com.

A CIP catalogue record for this book is available from the British Library

Hardback ISBN: 978-1-3985-3998-3
TPB ISBN: 978-1-3985-3999-0
eBook ISBN: 978-1-3985-4000-2

Typeset in Palatino LT Std by Palimpsest Book Production Limited, Falkirk, Stirlingshire

Printed and Bound in the UK using 100% Renewable Electricity at CPI Group (UK) Ltd

For my sons,
Zackaria and Salahdin

CONTENTS

Preface ix

Our Cosmic Contemplations

Chapter 1 The Ancient Cosmos 3
Chapter 2 The Copernican Revolution 17
Chapter 3 Pushing Boundaries 27

Our Cosmic Conquests

Chapter 4 Not Like Clockwork 1: Space and Time 43
Chapter 5 Not Like Clockwork 2: Overthrowing Gravity 57
Chapter 6 Building Blocks 1: Laws of Matter 73
Chapter 7 Building Blocks 2: Subatomic World 89
Chapter 8 Dark Universe 109

Our Search for Cosmic Company

Chapter 9 Life As We Know It 131

Chapter 10 Life in Our Neighbourhood and Beyond	153
Chapter 11 Decoding Messages from the Cosmos	179

Our Future in the Cosmos

Chapter 12 A New Era	199
Chapter 13 Our Cosmic Legacy	215
Epilogue	221
Acknowledgements	225
Bibliography	227
Index	239

A Brief History of the Universe

(and our place in it)

PREFACE

My family and I gather around a crackling fire under a moonlit sky studded with distant suns. Tucked away amid lush green mountains, with pine trees stretching skyward and snow-dusted peaks lining the horizon, our ancestral village of Mangloor in north Pakistan still bears intimations of the way we used to live. Here, days are bounded by the rising and setting sun. Life is attuned to the rhythms of nature; our intimate connection with the natural world is ever-present.

Far from the blinding lights and industrial din of urban life, the shimmering band of the Milky Way stretches overhead in high-definition splendour. It holds us in an attitude of reverent awe. We are dwarfed by its expanse, our brief tenure on this planet overshadowed by its ancient presence. It is not hard to see why humans have always looked to the heavens for meaning, petitioning them to give context to our existence so that we may come to know where we belong and how we fit into their grand design. We have been met with silence. We do not know whether we ask the right questions or if we will ever find the answers. But something in our nature compels us to rise above the frailty of our human condition and seek a union with that which will outlast us by an eternity.

Imprinted in our genes is the code needed to survive and thrive within the parameters of this planet. Yet it remains a marvel of human existence that we can comprehend worlds far removed from our own – from the fiery demise of faraway suns to the baffling complexities of the subatomic. Our partaking in a much grander and more mysterious reality may be among the greatest privileges life has to offer. It was this I sought to express in a series of letters to my young sons, letters that have grown into the book you now hold – an appeal to take in the magnificent wonder of the world, to apprehend the nature of the reality we inhabit and the universe that, in all its entirety, has brought about our genesis.

Before long, nature will dissolve our essence as she has for all those who came before us, reclaiming the atoms in our bodies to create countless other entities. Yet it may be of comfort to know that the totality of our being is inextricable from the totality of the universe. We are a part of this cosmos, and we have a place in that story.

Our Cosmic Contemplations

CHAPTER 1

The Ancient Cosmos

Under the silent, star-speckled darkness of an ancient night sky, our ancestors searched for the meaning of existence. Gazing into the depths of the cosmos, they wondered: what are those objects, and how do they relate to us? Following the trails of twinkling lights, they drew associations between what they observed overhead and the impact on their earthly lives, connecting the dots between the celestial and the terrestrial and tethering our ephemeral existence to the eternal theatre of the cosmos.

The stars served as their guides, charting paths for distant travels, dictating the sowing and reaping of crops and ushering in the rhythmic change of seasons. But the cosmos was far greater than a tool for navigation and timekeeping; it was also the abode of deities, steeped in myth, an eternal backdrop against which the fleeting epochs of humanity played out. It was with this mix of reverence and curiosity that we began our intimate relationship with the cosmos.

A BRIEF HISTORY OF THE UNIVERSE

BABYLONIAN ASTRONOMY

Around the seventh century BC, in the alluvial plains of Mesopotamia between the Tigris and Euphrates, the Babylonians watched the skies with a purpose that was both sacred and practical. Their night sky was filled with messages from the gods, who wielded influence over mortal affairs and expressed their will through the mechanics of celestial bodies. The heavens spoke, and deciphering this cosmic dialogue became a sacred obligation entrusted to the priests, who were not just religious leaders but also the chief astronomers and interpreters. Night after night, they scoured the skies and charted the positions of stars and other celestial phenomena, then meticulously etched them in cuneiform script on clay tablets.

Recorded in these 'Astronomical Diaries', which span from around the seventh century BC to the first century BC, are the detailed movements of the Sun, Moon and stars; lunar and solar eclipses; and accounts of the fiery trails of comets and the flashes of meteors. But interpreting the goings-on above also required keeping tabs on what was happening below. Interwoven with these celestial observations are commentaries on more worldly concerns: notes on the weather, the surging and receding of rivers and even market fluctuations in the price of staple goods.

The priests relied on both historical precedence and associative reasoning, such that, if a certain cosmic event had once preceded a period of drought or famine, its recurrence was a sure omen of similar hardships ahead. Similarly, a new moon appearing earlier than expected was a forewarning of troubles because of the negative connotations of premature events.

THE ANCIENT COSMOS

Astrology and astronomy were, in those days, one and the same. With seventy cuneiform tablets bearing around 7,000 celestial omens, it's palpably evident that Babylonian culture relied heavily on interpreting the skies. The lights up above commanded direct influence over both daily affairs and affairs of the state; they dictated the rise and fall of empires, warned of environmental catastrophes and even forecast the destinies of unborn children. Of all the heavenly oracles, though, the grandeur and precision of one body obscuring another was an especially potent sign: eclipses could be portents of doom, spelling the demise of kings. The Babylonians learned to be especially attentive to them.

Central to Babylonian astronomical practices was the sexagesimal (base-60) system of arithmetic notation. A scribe would use his stylus to mark a 1 by impressing it edgewise on a clay tablet and a 10 by pressing it flat, repeating this as needed to denote numbers from 1 to 59; but for 60 he would simply repeat the symbol for 1. With this innovation, large and complex numbers could be efficiently expressed without resorting to cumbersome notation, and there were no limits to the accuracy with which numbers could be represented. Astronomical observations could be recorded with unprecedented precision.

The Babylonians used this sophisticated system to trace the transit of planets. They partitioned the sky into 30-degree segments and to each they assigned one of the twelve zodiac signs, or constellations. As the planets traced their course along the ecliptic (the path the Sun appears to take across the sky), they would pass through these zodiacal signs, allowing their position to be accurately noted. In carving up the cosmos this way while reckoning time in hours, minutes and seconds, the priests made the heavens more mathematically manageable. That we continue to define time and angles in base-60 notation is an enduring legacy of this ancient era.

The Babylonians etched tablet after tablet with valuable and accurate data, creating one of the most extensive data collection exercises in history. The sheer volume of these astronomical records and the sustained commitment over hundreds of years were unparalleled until the later Greek astronomers.

Perhaps the greatest legacy of Babylonian astronomy was their system for predicting the next lunar eclipse. Using the Saros cycle, they knew that for any given eclipse there would be a near-identical one in timing, duration and intensity almost exactly eighteen years, eleven days and eight hours later. This recognition – that stars and planets moved in regular, repeating patterns – was revelatory. Keeping extensive records of their movements over long periods of time had given the Babylonians a powerful new tool. They could forecast the positions of heavenly bodies with far greater accuracy than ever before, and this became an apparatus of power and control for the rulers.

Since lunar eclipses that determined the fate of kings could now be predicted, the Babylonians weren't entirely helpless in the face of divine displeasure. Impending disasters could be averted if known ahead of time. Hence, a stand-in king was appointed during the period when the reigning monarch's life was seen to be in danger, and this anointed surrogate absorbed the ire of the gods, safeguarding the life of the king. Once the danger had passed, the substitute was stripped of all his regalia and often consigned to 'go to his fate', in other words, executed; the omen thus became a self-fulfilling prophecy.

The Babylonians' predictive power over celestial phenomena marked humankind's first attempt at devising a systematic framework for understanding the cosmos. It would richly influence the celestial pursuits of later civilisations, from the Hellenic world and the medieval Islamic empire to the contemporary West.

ANCIENT GREEKS

Early studies of the cosmos paved the way for the fervent intellectual discourse that flourished in ancient Greece, with its highly sophisticated tradition of philosophical and scientific enquiry. Placing emphasis on reason and rational observation, the Greeks promoted an impartial study of the natural world, striving to explain phenomena without recourse to the supernatural. This was the era that produced towering intellectual figures who would dominate philosophical and scientific thought for millennia.

It is to the ancient Greeks that we owe the word 'cosmos', which derives from the Greek word for 'order' and 'ornament/decoration'. To them, beauty was synonymous with orderliness, and the elegant harmony with which the heavens operated epitomised this. Everything followed its prescribed course, never wayward, never faltering; the sun rose every morning to set again every evening, leaving none in any doubt that it would rise again the following morning. This dependability extended to the other heavenly bodies such that the constellations of stars in the firmament were as they had always been, across many epochs of humanity. Great kingdoms rose and fell, mighty empires conquered and crumbled, all the vanities and vagaries of humanity played out below heavens that remained unchanging.

Arguably, the most influential thinker in Western intellectual history was Aristotle. In the fourth century BC this student of the formidable Plato revolutionised virtually every arena of thought he engaged with. His works spanned such an extensive range of disciplines, from philosophy and logic to politics and metaphysics, that in today's terms they would fill fifty substantial volumes. For Aristotle, happiness was rooted in a life of the mind, a life immersed in understanding the essence of existence and the world we inhabit, the 'knowledge of things human and divine'. In this exalted pursuit, he aspired to imitate the gods, counselling that we ought 'not listen to those who urge us to think human thoughts since we are human, and mortal thoughts since we are mortal; rather, we should as far as possible immortalise ourselves'. In the deepest sense, Aristotle succeeded in that: the Aristotelian worldview endured for thousands of years. In his impassioned desire to transcend the perceived limitations of human intellect, he outlasted his gods.

For Aristotle, science was far from amassing an inventory of facts; instead, it was the glorious art of arranging and integrating those facts into a coherent account of the world. His systematic methodology, coupled with his profound philosophical enquiries, gave him a distinctly unified outlook, which a tenth-century Byzantine lexicon would describe as someone who was 'a scribe of nature, having dipped his pen in thought'.

When the ancient Greeks looked up at their night sky, they understood Earth to be the unmoving centre of the universe, with a celestial dome rotating around it. Since the position of most stars remained unchanged relative to their gaze, the Greeks called these 'fixed stars'. Those that wandered across the sky against this backdrop of fixed stars they called 'planets'.

Aristotle duly noted that during an eclipse the Earth casts a circular shadow on the lunar surface, and this must mean that our planet is a sphere. His cosmological system thus posited a spherical Earth at the centre of a spherical universe. Beyond Earth lay the Moon, the Sun, the planets and the fixed stars, with each heavenly body attached to a concentric sphere. These spheres and the bodies they carried were perfect and incorruptible; they moved in circles, the embodiment of perfection, and were made of a special element (aether) that was divine and distinct from the four earthly elements (earth, water, air and fire). Unlike the mortal realm, which was subject to decay and change, the heavens were eternal and unchanging.

Aristotle thought that the physical universe was vast but spatially finite. Outside of the bounded sphere of the cosmos, he believed there was nothing – no space, no time, no void, no 'beyond'. Time, however, he considered to be eternal; the universe had always existed and would continue to exist; it had no temporal beginning nor an end. Beyond all this was the supreme cause of all motion: the 'unmoved mover'. For the heavenly bodies to be swept along in their perfect spheres, and for birth and decay to permeate all earthly entities, there must be a cause or source. The 'unmoved mover' is incorporeal and eternal, not chained to the confines of space or the constraints of time, existing outside the universe.

The spherical universe centred on the Earth encapsulated humanity's thoughts on the cosmos for many civilisations to come. The young prince Alexander of Macedon, whom Aristotle had tutored, became Alexander the Great and conquered much of the known world. Thus the Aristotelian worldview, along with the broader Greek astronomical tradition, was exported far and wide and, in its travels, encountered and assimilated with the rich observational astronomy of the Babylonians.

A BRIEF HISTORY OF THE UNIVERSE

Around AD 150, Ptolemy expanded on the geocentric view with an elaborate mathematical model to explain celestial movements, drawing in part on the data inscribed in the clay tablets the Babylonians had laboured over. A stationary Earth sat at the universe's centre, while the Sun, Moon, stars and planets orbited it in complex patterns. It was the motion of the planets that Ptolemy found especially baffling; they appeared to move in the same direction as the Sun (prograde motion), then halt and reverse their direction (retrograde motion). To account for this apparent zigzag motion, he introduced small circles that carried the planets (epicycles), which themselves moved on larger circles (deferents) around Earth. Although it wasn't understood why they moved in this way, this description provided a mathematically consistent method to predict planetary positions.

Ptolemy's seminal work, known as *Almagest* (derived from the Arabic *al-mageisti*, or 'great work'), presented a detailed account of this worldview, distilling centuries of astronomical observations. It would remain the authoritative textbook on astronomy until the Renaissance. An epigraph in the Latin translation states: 'I know that I am mortal by nature, and ephemeral, but when I trace at my pleasure the windings to and fro of the heavenly bodies I no longer touch the earth with my feet: I stand in the presence of Zeus himself and take my fill of ambrosia.' We cannot say for certain who wrote these words, yet the sense of exaltation they convey is undimmed. The ancients experienced the rapture of contemplating the vaulted realm, and those moments of apprehension were enough to lift them so far above the confines of the human condition that they communed with the divine and supped with the gods. The sky has never lost this hold over us.

THE ANCIENT COSMOS

How early civilisations viewed the cosmos is something we are continuing to piece together from fragments of what they left behind. In 1901, off the coast of the Greek island Antikythera, sponge divers stumbled on the submerged ruins of a shipwreck and recovered a curious artefact hiding under centuries of ocean deposits. This relic was later identified as the Antikythera mechanism, dated to around the first century BC – a highly intricate mechanical device that would have been used as a hand-powered model of the geocentric solar system. It was operated by turning a hand crank, which drove a series of interlocking gears that, in turn, moved several dials and pointers on the instrument's face. Astonishingly, this miniature assembly tracked the positions of the Sun and Moon, the phases of the Moon, and maybe even the movements of the planets along the ecliptic – the path they appear to follow across the sky.

In executing complex calculations to predict celestial positions, the Greeks were effectively tinkering with an early form of programmability. This ancient analogue computer hints at an astonishing level of sophistication – nothing comparable appears in the historical records for well over a millennium. Not only did the Greeks possess an advanced understanding of celestial dynamics, but they also had the technical know-how to apply this knowledge in a practical device.

The Greeks distilled their astronomical knowledge into treatises which were then translated into Arabic in the Arab-Muslim empires of the Middle Ages and widely disseminated – this would be the primary route via which ancient knowledge was preserved. Through this cross-cultural transmission, the Greeks' view of the cosmos survived the vicissitudes of history to make its mark on later astronomical traditions.

A BRIEF HISTORY OF THE UNIVERSE

ISLAMIC GOLDEN AGE

As the sixth century dawned, the fall of the Roman Empire plunged Europe into the shadows of the Middle Ages, marking a period of relative decline in scholarship across the continent. By contrast, the eighth to the thirteenth century marked the 'Golden Age' for the Islamic Empire. Stretching from the Iberian Peninsula in the west to the Indus river in the east, this burgeoning empire encompassed much of the Middle East and even parts of Central Asia. Here in the Islamic world, the astronomical wisdom of the Greeks, which had, in turn, learned from Babylonian discoveries, found its new custodians.

This was a culture of assimilating knowledge and preserving the scientific wisdom of civilisations that had been conquered. Many classic works of antiquity that might otherwise have been lost were translated from Greek, Syriac, Middle Persian and Sanskrit into Arabic, then copied and dispatched across the empire. Dubbed 'the translation movement' (Harakat al-Tarjama), this large, well-funded and sustained effort employed translators from diverse backgrounds and religious persuasions to develop an expansive inventory of educational literature. The House of Wisdom (Bayt al-Ḥikmah) in Baghdad became the beacon of this intellectual activity.

Knowledge of the cosmos flourished in this vibrant landscape. The cultural and scientific hubs of Baghdad, Damascus and Samarkand became the seats for advancing the study of the skies. Grand observatories were founded; the first in Baghdad and later ones around present-day Iraq and Iran. Devices like the astrolabe (Greek for 'star-taker') found prominence across fields as diverse as astronomy, religion, navigation and time-keeping. Among those skilled in making them was al-'Ijliyyah (now popularly known as Mariam al-Astrulabi), who apprenticed alongside her father under a renowned astrolabe maker in Baghdad. In addition to such portable tools, astronomers devised observational sextants, some extending up to 40 metres long, which were essential for deducing the angle of the Sun and the movement of the stars and planets. Scholars also advanced the maths underpinning these observations; Al-Khwarizmi established algebra as a distinct discipline, systematising its study and introducing new concepts and techniques, and Al-Battani extended the tables and formulas for spherical trigonometry, critical for calculating the positions and movements of celestial bodies on the celestial sphere.

Islamic scholars refined the solar and lunar models to greater accuracy than ever before, driven in part by the rituals of Islam. Practising the faith required knowing the precise times for daily prayers, the timing of dawn and dusk for the fasting month of Ramadan and the direction of Mecca for prayer and pilgrimage. So astrolabes were refined, new calendars were created and observational methods were improved. Intriguingly, the perceived conflict between religious faith and rational enquiry so entrenched in our modern narrative has not been a permanent fixture of human history. Science and spirituality can both lead to the sublime, and there have been extensive periods when both were not just allowed but even encouraged to coexist. This was one such period.

An important legacy of this era was the rigorous and methodical approach to scientific enquiry itself, championed by Ibn al-Haytham (Latinised as Alhazen). His revolutionary magnum opus on the science of light and vision, *Kitāb al-Manāzir* or Optics, was replete with systematically designed experiments and geometrical proofs that would go on to influence seventeenth-century luminaries such as Kepler, Descartes and Huygens. In his later treatise *Shukūk 'alā Baṭlamyūs* or Doubts on Ptolemy, written in the eleventh century, he laid bare some of the contradictions he found in Ptolemy's *Almagest* and stressed the need to critically evaluate existing theories and harbour a healthy scepticism. He argued that hypotheses must be supported by logic and observation and that neither science nor those who practise it are immune from error. Thus, a scholar must challenge the readings of other scientists and 'attack it from every side', but also 'suspect himself' and question his own biases and assumptions. In so doing, Al-Haytham had planted an early seed in the development of what we now recognise as the modern scientific method. It was our slow uncovering of this new way of thinking that facilitated a move beyond intuition and superstition to a more reasoned understanding of the world. Ultimately, this capacity for rational enquiry enabled us to challenge our long-held assumptions about the cosmos.

Islamic astronomers critically examined the Ptolemaic geocentric model and found some discrepancies with their own observations. The Ptolemaic model had an intricate system of epicycles and deferents that came to be seen both as overly complex and insufficient for accurately explaining celestial phenomena. Back in the ninth century AD Al-Battani had pinpointed these inaccuracies and refined the lengths of the year and eclipse predictions. Subsequently, in the thirteenth century AD, Al-Tusi developed a mathematical technique, the Tusi couple, which he used to replace Ptolemy's problematic equant for many planets. By showing that complex celestial movements could be explained without resorting to equants, he challenged a key component of the geocentric model.

THE ANCIENT COSMOS

This golden age of celestial enquiry eventually drew to a close. A constellation of factors was to blame: political fragmentation, economic challenges, the Mongol invasions bringing the destruction of libraries and madrasas, and internal strife within the Islamic world. With dwindling support for scientific endeavours and a discernible shift from rationalism to tradition and orthodoxy, the inclusive and freethinking religion that had nurtured a coexistence of intellectualism and belief was now profoundly changed.

Having preserved the wisdom of the ancients and advanced our thinking of the cosmos, the legacy of the golden era began to trickle into Europe, setting the stage for a new era of exploration. And while the Earth-centred narrative remained, the process of refining and rigorously examining that framework had exposed some cracks in the model, the unpicking of which would later lead to its unravelling.

CHAPTER 2

The Copernican Revolution

Western Europe was emerging from the Middle Ages with a renewed vigour for learning – one that would shake the very core of long-held cosmological views. From the fourteenth to the sixteenth century, the Renaissance, which originated in Italy before spreading across the continent, saw an intense resurgence of intellectual, cultural and artistic traditions, and a keen interest in the revival of classical antiquity. Scholars pored over the texts and philosophies of ancient kingdoms, and it was this spirited enquiry that lit the fuse for challenging entrenched scientific doctrines.

The prevailing model of the cosmos at the beginning of the sixteenth century was still intuitively aligned with the experience of gazing up at the heavens and watching them wheel overhead in a silent, majestic arc. From this vantage, seeing the Sun's daily journey across the sky, the Moon's glide through the night and the procession of the stars, the Earth appears to be a fixed anchor in a sea of celestial motion. But though it had dominated human thought for over a millennium, the geocentric universe was an illusion.

The first big reckoning for this centre-of-the-universe narrative came in 1543 when Polish scholar Nicolas Copernicus put forward a radically new hypothesis of the cosmos. A true polymath of the Renaissance era, Copernicus's intellectual pursuits spanned astronomy, medicine, economics and even Church law. He practised as a physician and served as a canon lawyer, while also delving into economic theories on the value of money. Yet, his enduring legacy came from his forays into astronomy, which would forever alter our view of the world.

Copernicus concluded that many issues with the Ptolemaic model could be elegantly resolved by positioning the Sun at the centre of the universe. Around this motionless sun, he arranged the known planets in their correct order, each tracing its path in a distinct orbital sphere, with the fixed stars occupying the outermost sphere. The apparent movement of the stars, formerly attributed to a rotating universe around a stationary Earth, was now explained by the Earth's own rotation. In this revolutionary model, our planet was no longer immobile; it orbited annually around the Sun, spun daily on a tilted axis, and a third 'motion of inclination' kept the axis nearly parallel through the year.

With the planets performing circuits around the Sun, Copernicus's model offered an elegant solution for the zigzag motion that had so perplexed Ptolemy. In Copernicus's heliocentric (Sun-centred) scheme, the apparent retrograde motion is due to Earth overtaking a slower moving outer planet, so that the planet appears to move backward against the stars. The planets aren't reversing their direction at all – their backward movement is an optical illusion.

THE COPERNICAN REVOLUTION

Copernicus regarded geocentric models of the cosmos as akin to grotesque disfigurations of the human form, with arms, legs and torso all mangled to resemble a monster. In contrast, his heliocentric vision preserved the integrity and harmony of the human figure.

Nature's manifest elegance, its regal simplicity, can help us gauge just how close to reality our theories of the universe are; whenever we've encountered inelegance, we've often concluded that we are some distance from the truth. Copernicus was led by the mathematical elegance of the Sun-centred universe.

He wasn't the first to question the geocentric model; the Greek Aristarchus of Samos had proposed a heliocentric model as early as the third century BC, and Seleucus of Seleucia, in the second century BC, also argued that the Earth orbited the Sun. But these ideas didn't find the cosmological consensus to become the dominant paradigm; with the authoritative backing of Aristotle, the Earth-centred universe prevailed. And although Copernicus moved away from the geocentric model, he retained the classical idea that the planets' orbits must be circular and uniform. Constrained by the belief that heavenly movements must be governed by perfect circles, he still used epicycles to account for the irregularities in planetary motion.

Copernicus initially shared his ideas in the 1510s but only fully elaborated on them in his seminal work, *On the Revolutions of the Celestial Spheres*, published in 1543, the year of his death. Copies seem to have been quite widely circulated, but the work didn't immediately spark the uproar we now associate with its reception, and Copernicus's ideas didn't upend our understanding of the universe until much later.

There were several reasons for this. While elegant, the model was not necessarily easier to use and didn't initially make predictions more accurately than the Ptolemaic one. More importantly, concrete observational evidence was lacking: the maths had got there before everything else. That 'the book of nature is written in the language of mathematics' would prove to be apt here, as it has been many times since, but without supporting observations such a monumental shift in cosmic perspective was going to be difficult to accept.

Even notable astronomers such as Tycho Brahe acknowledged the model's mathematical beauty and adopted some of its features, but continued to place a stationary Earth at the centre. Tycho was dubious that this 'hulking, lazy body, unfit for motion' could engage in the movement we normally associate with lighter, aethereal bodies, let alone possess a 'triple motion'.

Even today, the reality that we are stationed on a planet spinning at 1,000 miles an hour and hurtling around the Sun at nearly 70,000 miles an hour is a disorienting one. Our intuition tells us we are on solid, unmoving ground. With no external points of reference, Earth's constant, multiple movements remain imperceptible to us. Lacking conclusive evidence and challenging the long-prevailing worldview, while also at variance with intuition and certain biblical interpretations, the heliocentric model was going to have an uphill battle for acceptance. But the spark for the Copernican Revolution had been lit.

During the early seventeenth century, the efforts of astronomers like Johannes Kepler, with his three laws of planetary motion, further refined Copernicus's vision. Kepler was convinced that mathematics was the route to understanding the cosmos. He viewed geometry as the language of God, believing that studying it and the mathematical rules that governed overhead phenomena was akin to communing with the mind of the divine. While the Ptolemaic and Copernican systems had clung to circular orbits as a manifestation of celestial perfection, Kepler gave the planets elliptical orbits (first law). This simplified the picture of how planets moved, eliminating the cumbersome epicycles that had cluttered previous models.

Kepler's second law explained why planets appeared to move faster in the sky when they were closer to the Sun and slower when they were farther away, resolving the apparent irregularities in planetary speed with a universal principle. And his third law established the relationship between the time a planet takes to orbit the Sun and its distance from it. Together, these laws endorsed the heliocentric model with a precision that the geocentric system could not achieve.

Through these efforts, the Sun-centred universe was on firmer theoretical grounds. This dramatic repositioning of Earth – and, by extension, us – marked a paradigm shift in astronomy that also had far-reaching implications for philosophy and religion. As the German writer Johann Wolfgang von Goethe summarised some centuries later: 'Of all discoveries and opinions, none may have exerted a greater effect on the human spirit than the doctrine of Copernicus. The world had scarcely become known as round and complete in itself when it was asked to waive the tremendous privilege of being the centre of the universe.'

Ripples became tidal waves felt not just in the study of celestial mechanics but deep in the core of a species that had grown up in a different reality. Dethroned from our long-held centrality in the cosmos, our notions of human exceptionalism were thoroughly challenged, bringing to a head one of the most well-known historical confrontations between science and religion. To become accepted as the predominant worldview, this extraordinary claim of heliocentrism would require indisputable, observational evidence. And that came from an Italian astronomer and his diligent observations of our largest planetary entity.

GALILEO

In 1610, Galileo Galilei was experimenting with his homemade telescope, a device that, until then, had been used primarily for terrestrial rather than celestial purposes. Galileo's first telescope was a modest instrument, magnifying objects to three times, but with more experimenting he managed to enhance his device's power, first to eight times and then to an astounding thirty times. It was with this improvised tool that he would forever change our understanding of the cosmos.

One evening, he turned his scope towards the heavens and, peering down the lens at the giant of our solar system, Jupiter, he noticed what appeared to be three small, seemingly stationary stars hovering nearby. Over subsequent nights, the positions of these 'stars' shifted in relation to Jupiter, though they never seemed to depart very far from the planet, with one of them even vanishing behind it. This dynamic hide-and-seek of objects that were supposed to be 'fixed' led Galileo to an astonishing conclusion: these were not stars at all but moons in orbit around Jupiter. Jupiter had its own moons. He went on to identify a fourth moon, completing his discovery of what are now known as the Galilean moons: Io, Europa, Ganymede and Callisto, Jupiter's four largest satellites. This moon count has since skyrocketed to a staggering ninety-five – in fact, they've become so numerous that we've stopped naming them. But Galileo's discovery held critical significance. This was the first direct observation of a celestial body orbiting another celestial body that wasn't the Earth.

Other observations further reinforced the heliocentric model. When Galileo turned his attention to the brightest planet in the night sky, Venus, he noticed that it exhibited to us a full cycle of phases, not dissimilar to the waxing and waning of our moon. He could reconcile this with Venus orbiting around the Sun and the changing angle at which sunlight illuminated the planet as it did so.

Repeated observations of the Moon revealed a landscape of bright protrusions and shadowy depressions that Galileo thought were much 'like the face of the Earth itself, which is marked here and there with chains of mountains and depths of valleys', while the Sun was similarly blemished with discoloured sunspots. In other words, these were not the flawless heavenly bodies that Aristotelian cosmology had long proclaimed.

This entire body of evidence stacked up such an extraordinary case against the prevailing worldview that the alternative became impossible to ignore.

What followed was intense scepticism and controversy. So deeply entrenched was the Earth-centric perspective that many simply refused to believe Galileo's findings: that Earth was just another celestial body; that humankind was not at the centre of the world, as literal interpretations of religious texts would have us believe; that the heavens observed wheeling across the sky in their apparent orbit around the Earth were a mere optical illusion. This was a cataclysmic upheaval in how we viewed the universe and ourselves. It caused a titanic clash of ideologies, bringing to the surface the simmering tension between emerging scientific enquiry and established religious doctrine.

A BRIEF HISTORY OF THE UNIVERSE

The political climate in the late sixteenth century was especially grim; the Protestant Reformation had left a lasting imprint on the Catholic Church, and the Church leadership was desperately seeking to reassert its authority. The moderate theology of Thomas Aquinas and the natural philosophy espoused by Aristotle was adapted into a rigid orthodoxy that left little room for alternative viewpoints. During this turbulent period, the Church established the first official Index of Prohibited Books and formed the Congregation of the Index to oversee its censorship efforts. A wave of condemnations followed – publicly denouncing those whose works were considered heretical or contrary to Church doctrine – making it perilously risky to question the prevailing orthodoxy. In the context of this political and religious tension, scientific debates were not confined to scholarly circles but tied to broader ideological and political conflicts. This context is critical for understanding the reception of new scientific ideas and the changing stance of the Church over the years of the Galileo affair.

The Catholic Church in the early 1600s was aligned with the Aristotelian geocentric perspective. The accepted scientific position was, therefore, also the theological one. After he published his findings in *The Sidereal Messenger* (a brief treatise on his telescopic observations), Galileo's fame and influence grew. He began openly advocating for the Copernican system, arguing that it was not only scientifically valid but also not contrary to scripture. While the Church's initial reaction was cautious, it grew increasingly hostile. In 1616, it declared the Sun-centred cosmos to be 'formally heretical' – that is to say, in direct contradiction to the divine word. Galileo was warned not to hold, teach, or defend heliocentric ideas. That same year, Copernicus's seminal work was also banned and only authorised for republication after the text was modified to present heliocentrism as merely a hypothesis.

Galileo did not initially see his findings as contradictory to religious beliefs. His observations had provided tangible proof that not everything in the heavens revolved around Earth. But he believed that, rather than contradicting the Bible, these discoveries simply required a different interpretation of certain biblical passages. In his famous letter to the Grand Duchess Christina – mother of the ruling Duke of Tuscany and a patron of the arts – he emphasised that there is no inherent conflict between the truths revealed by scripture and those discovered through scientific enquiry, that the book of nature and the Bible were essentially different 'books' written in different 'languages' – but God was the author of both and 'two truths cannot contradict each other'. He argued that theology and science should remain distinct because they address different realms of knowledge; people turned to the Bible for spiritual guidance and to nature for understanding the physical world.

In 1632, Galileo published *Dialogue Concerning the Two Chief World Systems*, presenting arguments for both the Ptolemaic geocentric model and the Copernican heliocentric model, with the dialogue heavily favouring the latter. This was seen as a direct challenge to Church doctrine and led to his trial by the Roman Inquisition in 1633, during which he was found 'vehemently suspect of heresy' – a less severe verdict than outright heresy. He was forced to recant his support for heliocentrism and placed under house arrest, where he remained until his death in 1642. In 1992, Pope John Paul II, following a lengthy investigation by the Pontifical Academy of Sciences, acknowledged that the Church had erred in its handling of the Galileo affair.

A BRIEF HISTORY OF THE UNIVERSE

As for Copernicus, he rested in obscurity for hundreds of years after his death, having been buried in an unmarked grave at Frombork Cathedral in Poland, where he had served as a canon. It wasn't until 2004 that Polish archaeologists used historical records and radar to exhume what they suspected might be his remains. Without a DNA sample from any relatives of Copernicus they had no conclusive proof – until Copernicus's personal library, preserved in Sweden, furnished several strands of hair that had been caught in the pages of one of his books. Subsequent DNA analysis confirmed two of these hairs to be a match with the exhumed remains, and in 2010, nearly 467 years after his death, Copernicus was given a ceremonial burial and a black granite tombstone emblazoned with a golden sun encircled by its planets – the Sun-centred system he had envisaged and that we have inherited.

The legacy of this era reverberated for a long time. Science was casting off the vestiges of entrenched narratives and stepping into the light of empiricism, carving a path for itself that was increasingly divergent from that of religion. The search for the ultimate truth of the cosmos ousted humankind from an exalted position at the centre of the universe, one that was supremely endorsed by religious doctrines. In doing so, it left a vacuum. Where once there had been some assurance of our place and purpose as sanctioned by the divine, we were now relegated to relative obscurity. Perhaps most unsettling was not the dramatic upheaval in worldview but the realisation that there *could* be such grand upheavals; that the reality we inhabited could be upended by the next breakthrough in favour of a radically alternative paradigm. The pursuit of rational understanding, while it demystified the heavens, also deepened our existential enquiry.

CHAPTER 3

Pushing Boundaries

The Copernican revolution had set the stage for a transformative era of enquiry: the Scientific Revolution, spanning the period from the mid-sixteenth to the late seventeenth century. During this time, the scientific method was fully embraced, a systematic approach to discovery that elevated the value of empirical evidence and rational analysis. We honed our tools, refining telescopes and microscopes to peer further into the realms of the macro and the micro. We also established esteemed scientific societies such as the Royal Society in London (1660) and the Academy of Sciences in Paris (1666), thus offering crucial institutional support for the burgeoning scientific community.

It was in the midst of the new age of enquiry that Isaac Newton made an indelible contribution to humanity's understanding of the universe.

Born on Christmas Day in 1642 in an English village, Newton apparently weighed less than two pounds and was small enough to 'fit into a quart pot'. Infant mortality was high and post-natal care rudimentary, and so the prognosis for such babies was haplessly poor – little Isaac wasn't expected to survive for long. But that was only the beginning of his troubled childhood. His father had died three months before he was born, and his mother remarried three years later, going to live with her new husband and leaving her son in the care of his grandparents. He was a sober and solitary child who spent his time reading and daydreaming and building extraordinarily accurate prototypes of windmills and wooden clocks and complex sundials. After his mother returned following the death of her second husband, she took him from school, hoping that he would one day take his place presiding over the family estate. It soon became apparent to his uncle and schoolmaster that Newton was most unsuited to such a life, and they persuaded his hesitant mother to let him return to school in preparation for Cambridge University.

In 1661, Newton enrolled at Trinity College Cambridge, the most prestigious college in England, where he would continue to cut a lonely and dejected figure, so consumed by his work that, according to the person he shared a room with, he would forget to eat or sleep. Just as he finished his degree at Cambridge, the Great Plague of 1665 ravaged England. This last major outbreak of the bubonic plague wiped out around 15 per cent of London's population and caused widespread panic and shutdowns. Cambridge also closed its doors to stem the spread of the deadly disease, forcing Newton to retreat to his family's country estate in Woolsthorpe. It was here that he at last found the safety and solitude to conduct the studies that laid the groundwork for modern physics.

During this forced isolation, Newton embarked on what he later referred to as his 'year of wonders'. The quiet setting of the English countryside, away from the distractions and duties of academic life, enabled him to contemplate the wonders of the natural world, and this period became the most productive of his life.

GRAVITY

In arguably the most famous tale in the history of modern science, Newton was at this farmhouse estate in deep thought when he saw an apple fall from a tree in the orchard. It was apparently this that prompted him to wonder why objects always fall downwards, and if this same force could extend beyond the Earth to influence the motion of the Moon and other celestial bodies. Pursuing this line of thinking allowed him to connect seemingly unrelated phenomena, ultimately leading him to formulate the laws of universal gravitation. Newton's revolutionary insight was that a single, universal force was acting on all objects with mass; the apple falling to the ground and the Moon sweeping around the Earth were both a consequence of this force. Gravity unified the mechanics of the celestial and the terrestrial under a single theoretical umbrella.

Newton's law of universal gravitation posited that every object in the universe attracts every other object with a force that is directly proportional to the product of their masses. The more massive an object is, the stronger the gravitational pull it exerts on other objects around it. This attractive force weakens as the distance between the objects increases, falling as the square of the distance separating them.

This was a revolutionary insight. Our understanding of why things fall to Earth was inherited mainly from Aristotle, who believed that objects fell towards their natural place, which for earthly objects was the centre of the Earth. Newton's gravity, by contrast, posited a single phenomenon to account not only for the dynamics on Earth, but also for those that transpired above us.

Our now well-established model of the cosmos had all the planets orbiting the Sun and the Earth spinning on its axis. Thanks to Kepler, we knew that planets orbited in elliptical rather than circular paths, that they moved faster when closer to the Sun than further away, and that there existed a relationship between the orbital period of a planet and its distance from the Sun. We knew quite well how planets moved, but not why they moved the way they did. Kepler's laws hinted at some underlying principle, and it would need Newtonian gravity to explain the why of planetary motion. It was this force that was choreographing the motions of celestial bodies.

While the invisible force of gravity elegantly explained observed phenomena, Newton was notably uncomfortable with the implications of 'action at a distance'. That objects could exert a force on each other across large distances and through a vacuum, without any mediating substance, struck him as so 'great an absurdity' that he refrained from proposing a specific mechanism and acknowledged it as owing to 'causes hitherto unknown'.

Newton presented his laws of motion and theory of universal gravitation in his magnum opus, *Principia*, which became the cornerstone of modern science. It not only advanced our understanding of the universe but also encouraged a more rigorous, mathematical approach to that understanding, a methodology that has remained a hallmark of scientific enquiry.

Despite his revolutionary insights, in a later memoir Newton reflected with great humility on his life's work:

PUSHING BOUNDARIES

> I do not know what I may appear to the world, but to myself I seem to have been only like a boy playing on the sea-shore, and diverting myself in now and then finding a smoother pebble or a prettier shell than ordinary, whilst the great ocean of truth lay all undiscovered before me.

Newton's words would come to ring true even for the magnificent theory that he proposed. The next tide to wash onto the shores of his 'great ocean of truth' would fundamentally redefine our understanding of gravity. Newton's revolutionary theory would prove to be the precursor to more profound mysteries of the cosmos, where even our notions of fixed entities like space and time would be overturned.

Meanwhile, Newton's laws were indicative of the era's move towards reason. His arrival at a descriptive framework through precise mathematical laws exemplified the potential of the new science to reveal the workings of the natural world. As natural philosophy evolved, it increasingly intersected with theological arguments, especially among clergymen in England who were also natural philosophers and eager to exercise this new empiricism to substantiate the existence of God. Rather than relying purely on biblical scripture or philosophical arguments, there was a bid to purge Christianity of its superstitions and bring it into the modern light of scientific enquiry; to use reason to reinforce religion.

Some viewed Newton's discoveries as powerful evidence of a divinely ordered universe. Far from a chaotic assembly, the universe appeared to be a finely tuned system governed by intelligible laws that were seen as the handiwork of a supreme 'Lawgiver'. Newton himself thought 'this most beautiful system of the sun, planets, and comets could only proceed from the counsel and dominion of an intelligent and powerful Being' – a divine power skilled in mechanics and geometry.

Newton's laws showed us how a set of mathematical equations could explain the motions of all celestial and terrestrial objects. The heavens that we had always seen as remote, operating on principles unlike those that governed terrestrial entities, were not so distinct after all. Earth may not be the centre of the universe, but we could now apprehend how the universe operated, and that did much to lift us above the insignificance of our place in it. It was this sentiment that French mathematician and physicist Blaise Pascal expressed in his popular classic *Pensées* (published posthumously in 1670): 'By space the universe encompasses me and swallows me up like an atom; by thought I comprehend the whole world.'

MORE PLANETS

With Newton's laws, we had not only a new way to see the universe and its underlying dynamics but also a powerful new toolkit with which to explore them further. They proved to be the gateway to remarkable breakthroughs.

The English astronomer Edmund Halley used Newtonian principles to calculate the orbit of the comet he observed in 1682, which bears his name. Sifting through historical records, he noticed that comets appearing in 1531 and 1607 had very similar orbits and wondered if they were in fact the same comet returning at regular intervals. Using Newton's equations, Halley calculated the gravitational tug of the Sun and planets on the comet's orbit and predicted that it should reappear in 1758. Although Halley was not alive to see it (he died in 1742), the comet's blazing return was a landmark validation of Newton's theory and a major leap for predictive astronomy.

In 1781, William Herschel, a musician and self-taught astronomer, was surveying the sky using a homemade telescope in his garden in Bath, England. Born in Hanover, Herschel had moved to England as a young man and built a career as a composer and organist before being drawn to the stars. One evening, he spotted an object that appeared to move against the fixed stars; it had a trajectory that led him to suspect it might be a new comet. In reporting his observations to the Royal Society, he noted that the object was much larger and less luminous than expected of any comet and with a body that was 'very well defined having neither beard nor tail'. Over time, calculations of this object's puzzling path revealed it was not a comet at all but a new planet, far beyond the orbit of Saturn, then at the known edge of the solar system. Herchel, using just his garden telescope, had discovered Uranus. At a time when Britain was embroiled in the American Revolutionary War and had lost thirteen of its American colonies, this gaining of ground in the heavens was, for some, akin to the conquest of a new realm and offered consolation. The physician and lecturer Matthew Turner wrote: 'It is true we had lost the *terra firma* of the Thirteen Colonies in America, but we ought to be satisfied with having gained in return by the generalship of Dr Herschel a *terra incognita* of much greater extent in *nubibus* [in the heavens].'

Labouring alongside Herschel in his garden observatory, and throughout his astronomical pursuits, was his sister, Caroline. Brother and sister were an inseparable team, devoting decades to tirelessly 'minding the heavens'. Herschel would observe through the night and dictate his findings to Caroline, who, the following morning, would calculate the positions of the observed objects and catalogue them. 'Every moment after daylight was allotted to observing', she would later recall, and 'if it had not been sometimes for the intervention of a cloudy or moonlight night, I know not when my brother (or I either), should have got any sleep'. In 1783, Herschel gave his sister her own telescope. 'I found I was to be trained for an assistant-astronomer', she wrote, 'and by way of encouragement a telescope adapted for "sweeping" . . . was given me. I was to sweep for comets.' Between 1786 and 1797, she discovered eight of them, including 35P/Herschel–Rigollet, a periodic comet which is expected to return to our skies in 2092.

After her brother's death, a distraught Caroline moved back to Hanover but continued her astronomical labours, now collaborating with Herschel's son and her nephew, John Herschel, who would also become a renowned astronomer. She completed cataloguing the 2,500 nebulae and star clusters she had compiled with her brother, arranging them into zones based on similar polar distances. In doing so, she laid the groundwork for the later creation of the New General Catalogue (NGC), a system still in use today. For her notable contributions to astronomy, Caroline Herschel was conferred numerous distinctions, becoming the first woman to be awarded the Gold Medal by the Royal Astronomical Society and later receiving the Gold Medal for Science by the King of Prussia. Upon her death at the age of ninety-seven, an obituary published by the Royal Astronomical Society read:

> Her memory will live, with that of her brother, as long as astronomical records of the last and present century are preserved; and it will live on its own merits, even though,

as may reasonably be hoped, the time should come when the astronomical celebrity of a woman will not, by the mere circumstance of sex, be sufficient to excite the slightest remark.

With William Herschel's discovery of Uranus, our cosmic family had been sensationally extended; there were more planets than we'd realised in the gravitational grip of our Sun. Orbiting so far out, around twice the distance from the Sun as Saturn's, Uranus also effectively doubled the size of the known solar system.

Observations of the newly discovered planet continued for many decades. This monitoring showed things that weren't expected: the orbit predicted by Newtonian mechanics, and the planet's observed orbit, were inconsistent. These 'wobbles' in Uranus's orbit suggested that the forces acting on it were not fully accounted for by the known planets. They could, however, be explained if we added another, as yet unseen, planet to the mix. A more distant planet still, exerting its gravitational influence on Uranus, could reconcile these observations.

By 1845, Uranus had completed nearly one full revolution around the Sun since its discovery, and this presented astronomers with an extraordinary opportunity to test their theory. Independently, both Urbain Le Verrier in France and John Couch Adams in England performed long and laborious mathematical calculations to predict the location of this possible new planet, based on its supposed gravitational influence on Uranus. Le Verrier mailed the co-ordinates of the 'region of the heavens where, perhaps, there is a planet awaiting discovery' to German astronomer Johann Gottfried Galle at the Berlin Observatory. The letter reached Galle five days later, and that very night in September 1846 he made the sensational discovery of Neptune, which was, remarkably, only one degree away from where Le Verrier said it would be. The predictive power inherent in Newton's laws had led us to Neptune with astounding accuracy and earned Le Verrier the distinction of being 'the man who discovered a planet with the point of his pen'. With hindsight, it transpired that others had glimpsed Neptune long before, starting with the astute Galileo in 1612, but due to its slow motion against the backdrop of stars, they had failed to identify it as a planet.

While these newfound planets had extended our solar family, our cosmic perspective at this point in the mid-nineteenth century was still markedly contained. We had no reliable way to measure the distances to faraway stars. This made it nearly impossible to understand the true scale of the universe and the remoteness of star clusters and nebulae. Yet, we believed that our galaxy comprised the entirety of the universe. It was vast and teeming with stars, planets and nebulae, but still only a single, immense galaxy. The perspective was of a world that, despite its size, seemed comprehensible, with us nestled somewhere central within its grand architecture and with nothing beyond our galactic horizon. We were soon to be toppled from that pedestal yet again.

NOT THE ONLY GALAXY

In the early 1900s, working in obscurity at the Harvard College Observatory was Henrietta Swan Leavitt, one of a group of women known as the 'Harvard computers', hired to analyse the outpouring of astronomical data being collected through photographic plates. Leavitt was studying special types of stars called Cepheid variable stars, whose brightness changes over time. While examining those in the Small Magellanic Cloud (a diffuse cloud-like patch in the southern sky), she noticed a striking pattern: the longer a Cepheid star took to brighten and dim, the more luminous it was. This meant that the period of a star's pulsation was directly related to its true brightness, and Leavitt could extract a clear mathematical relationship between the two. These stars, it turned out, were excellent yardsticks for measuring cosmic distances. Leavitt's breakthrough gave astronomers a reliable means for estimating the distance to faraway stars and star systems, something that had long been one of astronomy's most difficult challenges.

In the 1910s, American astronomer Harlow Shapley was keen to understand just how vast the Milky Way is and where our solar family resides within it. Using a dense cluster of stars that orbit the core of our galaxy, Shapley started piecing together a map. Building on Leavitt's findings, he used Cepheid variable stars to calculate how far away the clusters were and discovered that they appeared to be mostly congregating around a point that was very distant. The centre of the Milky Way, and therefore its gravitational heart, was not near our Sun, but far from it. Our galaxy was more immense than we had thought, and the life-sustaining Sun that our ancestors had worshipped as a god was not in the star-packed centre but in a quiet and distant suburb on its periphery.

But was the Milky Way the entirety of the cosmos? At the Great Debate of April 1920, one of astronomy's defining moments, Harlow Shapley and Heber Curtis, two towering figures in the field, clashed over this most fundamental question. Shapley argued that our galaxy was all there was: a lone island of stars in an ocean of void. He insisted that the Milky

Way was so vast that it encompassed all the stars in the universe, with the Sun at its outskirts. The 'spiral nebulae' (spiral-shaped clouds of dust and gas) and objects like the Andromeda Nebula (the nearest of all the great nebulae), he believed, were parts of this massive galaxy. Curtis, however, challenged this view and contended that the Milky Way was not a singular entity but one of many. He pointed to the dimness of novae (exploding stars) in Andromeda as evidence that it was far beyond our galaxy. Those 'spiral nebulae' were not gas clouds within our Milky Way but distant galaxies in a universe filled with many such 'island universes'.

The debate didn't immediately resolve the mystery. With the tools available at the time, it was difficult to measure conclusively the vast distances to objects like the Andromeda Nebula. But thanks to the pioneering work of the American astronomer Edwin Hubble, the impasse was soon broken. In 1924, Hubble was stationed at the Mount Wilson Observatory in California and had at his disposal one of the most powerful optical telescopes in existence. He turned this 100-inch Hooker Telescope towards the Andromeda Nebula and, using the Cepheid variable stars as reliable rulers, studied those within Andromeda and measured their distance from Earth. His calculations offered a staggering conclusion: Andromeda was exceptionally far away – much too distant to be part of the Milky Way. It had to be an entirely separate galaxy. Hubble wrote to Shapley reporting what he had found. On reading the note, Shapley told a colleague, 'Here is the letter that destroyed my Universe.'

These nebulous stars had been observed for millennia. In 964, the Persian astronomer Al-Sufi described what he saw as 'a small cloud' in his *Book of Fixed Stars*, an account we now know to be the first recorded observation of the Andromeda galaxy. It was Hubble's work that conclusively showed how such nebulae were indeed independent galaxies, far beyond our own; that the dim, milky band of light stretching across our night sky was not the cosmos's sole galaxy.

But Hubble was not done yet. By 1929, he had discovered that these faraway galaxies were not stationary; rather, they were hurtling away from us – and from each other – at tremendous speeds. The universe was expanding. And the more distant the galaxy, the faster it was receding from us. The modest cosmos we once envisaged, centred around us, was unfolding into one of implausible magnitude; its newfound dynamism a great upset to the steadfast and unchanging firmament we had known since antiquity. Our telescopes compelled us to 'measure the world by its scale, and not ours' and to concede that life on the human plane was not going to prepare us for the worlds that lay incomparably beyond.

Our Cosmic Conquests

CHAPTER 4

NOT LIKE CLOCKWORK 1

SPACE AND TIME

By the end of the nineteenth century, after a series of dizzying leaps in how we viewed the cosmos, we were finally in possession of theories that described an array of phenomena with remarkable success. Newtonian mechanics gave us a framework capable of explaining the motion of terrestrial and celestial objects, from the falling of apples to the orbiting of planets. After Newton, we perceived a mechanistic universe that was vast and intricate, but largely understandable and predictable. Space was an unchanging, three-dimensional stage on which physical events occurred. Time was a similar but entirely independent entity, ticking away with absolute regularity, regardless of events or observers. And the motions of the stars and planets were choreographed by the invisible force of gravity.

There was perhaps some sentiment that major upheavals in our understanding of the cosmos were largely behind us. This tentative optimism was expressed in a speech in 1894 by

German American physicist Albert A. Michelson, who surmised: '. . . it seems probable that most of the grand underlying principles have been firmly established . . . An eminent physicist has remarked that the future truths of physical science are to be looked for in the sixth place of decimals.'

Soon after the turn of the century, however, sweeping discoveries shattered this illusion and dispelled any idea of being near the end in our search for the universe's underlying principles. The universe that had become predictable to us was not quite as it seemed. The deterministic, clockwork world of Newtonian mechanics gave way to a new reality, governed by probabilities and uncertainties at the quantum level and warped spacetime at the cosmic level, altering our views on the smallest and the largest scales. So grand have the revolutions of the past century been that, no matter how spectacularly successful our description of the universe, we remain perpetually mindful that the next wave of experiments could usher in yet more radical revisions.

Even as we were celebrating the triumphs of the nineteenth century, in 1900, Lord Kelvin, a British physicist best known for his contributions to thermodynamics, delivered his 'Two Clouds' speech at the Royal Institution. In it he singled out two challenges (or clouds) hanging over the received theories of heat and light. The first was that the mysterious substance called the 'luminiferous aether' – a hypothetical medium believed to permeate all of space and through which light waves propagated, in the same way that sound waves travel through air – had failed to show up in experiments. If there was no aether, how did light travel through the vacuum of space?

Secondly, he questioned our understanding of how energy must be distributed in a heated system. A discrepancy between our predictions and the way certain gases behaved in experiments hinted that we were missing something fundamental about the nature of matter – it wasn't adhering to the laws we'd ascribed to it.

Kelvin was optimistic that these 'two clouds' would clear away in the not-so-distant future. He was right: they *did* indeed dissipate, but not before they had unleashed such torrential downpours that when the skies cleared, the universe was unrecognisable from the classical worldview we had fostered for so long. The two major revolutions of relativity and quantum mechanics dismantled centuries of accumulated wisdom, so that the entire edifice of physics constructed up until that point is now referred to as 'classical physics'. It was a paradigm shift that demarcated a new era of our understanding of the cosmos.

It was with Newton that questions about the intrinsic nature of time and space had become an integral part of physics: the subjects of empirical investigation as well as of philosophical discourse. Newton's rival, the German philosopher Gottfried Wilhelm Leibniz, opposed this absolute perspective, arguing that space and time were not distinct entities but, rather, abstract frameworks that we used to represent the relationships between things. For Leibniz, space was not a pre-existing empty container in which events occurred; rather, it was how we understood the connections between objects – like a map that shows the relationships between locations without implying the existence of an underlying empty space. Time was similarly a relational construct, serving as a means to deduce the ordering of events. Within philosophical circles, a stalemate existed between these perspectives. But with the overwhelming success of Newtonian theories, it was the Newtonian views on the nature of space and time that ultimately prevailed.

MOTION IS RELATIVE

The fact that motion is relative can be a disorienting phenomenon we all encounter in our everyday experience. For example, when you feel your train pulling away from the platform, only to realise that it is, in fact, the adjacent train that is moving and not yours, or when you are in a plane smoothly cruising at hundreds of miles an hour, yet you sense that you are not moving at all. Without an external reference point like the station platform or the ground rushing by as you look outside, it is difficult to determine whether you are moving at all.

The idea that motion is relative has been with us for centuries. In 1632, in *Dialogue Concerning the Two Chief World Systems*, Galileo argued for the concept that the Earth's rotation would not be detectable through mechanical experiments conducted on Earth. He was trying to address the common objection that if the Earth were moving, we should feel its motion; objects falling from a height should land behind their point of release because the ground has moved in the interim. That the Earth's motion would not be apparent to us was a crucial point in Galileo's defence of the Copernican heliocentric model.

Galileo asked us to imagine ourselves sailing smoothly on a ship, situated below deck in a windowless cabin, where there are no external visual cues. Lacking outside reference, in this isolated environment, the ship's motion, or lack thereof, would be imperceptible through mechanical experiments conducted within. Whether the ship was anchored at harbour or sailing across seas, water droplets from a suspended bottle would drip vertically, butterflies would flutter freely in all directions, fish in a bowl would swim indifferently in all directions, and jumping up and down or throwing an object back and forth would yield identical results. In other words, Galileo asserted, there is no experiment you can do to prove you are in motion: the laws of mechanics hold in either case.

Hence, the distinction between 'moving' and 'stationary' is all relative, and we cannot discern the motion of the Earth simply by its effects on terrestrial phenomena. This is often referred to as Galilean relativity and it laid the groundwork for the emergence of the special theory of relativity in 1905, introduced to the world in a paper submitted by a young clerk, Albert Einstein, working in a Swiss patent office. That year, Einstein put forward not one but four extraordinary works. They exerted the most profound influence on the way we viewed the universe – as well as extensive discussion and conjecture about how an unassuming 26-year-old with no distinguished academic standing could do it.

EINSTEIN

While he seems to have shown an early fascination with the workings of the universe, the young Einstein was not considered a prodigy. He rebelled against the rigid school system in Germany, with its authoritarian teaching style and emphasis on rote learning over critical thinking. In 1895, at the age of sixteen, he left Germany and moved to Switzerland, where he enrolled in a progressive institution known for its more liberal and open teaching methods. Here, he lived with the family of Jost Winteler, a teacher at the school and with whom Einstein developed a close bond. The Winteler household was liberal, engaging in frank discussions over dinner, even about controversial topics. But while Einstein was happier in this environment, his school report card wasn't exactly glowing.

When a concerned Jost Winteler sent Einstein's results to his father, Hermann Einstein responded that 'not all its parts fulfil my wishes and expectations, but with Albert, I got used a very long time ago to finding not-so-good grades along with very good ones, and I am therefore not disconsolate about them'.

It was in Switzerland that 16-year-old Einstein said he indulged in his 'first juvenile thought experiment which has to do with the special theory of relativity'. He imagined what it would be like to chase a beam of light. He thought that if he could keep pace and ride the light wave like a surfer, then, according to the principles of physics, he should see the wave as frozen in place, so that it would appear to him as though stationary.

At the age of seventeen, Einstein enrolled in the Swiss Federal Polytechnic School (now ETH Zurich) to study maths and physics, where he continued his defiant relationship with the structured education system, shunning classes he thought irrelevant or uninteresting and opting instead to study on his own. He became close friends with fellow student Marcel Grossman, who introduced Einstein to differential geometry, a prescient development that would play an important role later in his theory of general relativity. It was also here that Einstein met his future wife and fellow physicist, Mileva Maric, one of the few women studying at the polytechnic.

After graduating in 1900, Einstein struggled to secure an academic position; his nonconformist attitude alienated professors and meant that he was unable to obtain the strong letters of recommendation necessary for an academic post. As the only graduate in his section not to secure a position, he lived a meagre existence for nearly two years before Grossman's father helped him secure a job at the Swiss Patent Office. There, he assessed patent applications for electromagnetic devices, a role that engaged his visual understanding of the world and, in his view, 'stimulated [him] to see the physical ramifications of theoretical concepts' – a critical skill for the conceptual leaps that would define his contributions to science.

On top of Einstein's professional insecurity there was family pressure to break up with his fiancée, whom he married only after his father's death in 1903. With a child, born a year later, came the financial strain of supporting a young family on his modest income as a patent clerk. As John Stachel writes, Einstein was not some 'tranquil academic, brooding at leisure on weighty intellectual problems' but, rather, a man facing professional and personal uncertainty.

SPECIAL RELATIVITY

It was under these grinding circumstances that, in 1905, Einstein published his landmark paper proposing what we now refer to as the theory of 'special relativity'. Taking ideas about relativity that we had held since the time of Galileo, he extended them to *all* of physics, including the nature of light. It had been established that the way objects move and forces interact (laws of mechanics) are indifferent to whether an observer is stationary or moving at a constant speed. But the nature of light had long been an outlier, set apart from this framework and posing unique challenges that had not yet been addressed.

In 1676, the Danish astronomer Ole Christensen Roemer, at the Paris Observatory, recorded the eclipses of Io, one of Jupiter's largest moons, as it passed behind the planet. He noted that the times at which the eclipses occurred differed depending on the distance between Earth and Jupiter (which varied as they orbited the Sun); sometimes, Io emerged from Jupiter's shadow a bit later than predicted, while at other times it was ahead of schedule. It was when Earth and Jupiter were further apart that the eclipses seemed to be delayed. Why should Jupiter's satellite take longer to come out of its shadow at these moments? Roemer reasoned that this must be because light from Jupiter's moon takes longer to reach us when we are farther away. Light, in other words, had a finite speed. This was the earliest recorded evidence; however, the implications for our understanding of space, time and motion would not be fully realised until Einstein.

We have now measured light's speed to be a staggering 300,000 kilometres per second. Fast though this may seem, the universe is vast and the figure quite modest given the distances involved. Sunlight takes around eight minutes to reach Earth; if our mighty star were to suddenly cease shining, we would remain oblivious until eight minutes later. By the same measure, when we gaze at the glowing embers in our night sky, we are looking into the very deep past, to an archive of the cosmos, to distant suns that might have long since been extinguished. The light we see from these stars might have left them long before human history even began, long before there was anyone here to bear witness, and our chronicling of their movements may be the only record of their existence.

An understanding of how light actually propagates didn't emerge until 1865, when British physicist James Clerk Maxwell suggested that, much like ripples on a pond, light was a disturbance in the electromagnetic field. Just as sound needs air or water for transmission, it was widely believed that light too requires a medium for its propagation. This mysterious medium, the luminiferous ('light-bearing') aether, was an invisible, undetectable substance thought to fill all of space. But the aether had stubbornly failed to show up in any of the experiments designed to discover its presence. This was one of the 'clouds' that Lord Kelvin saw as obscuring the elegance of nineteenth-century physics – the seeming absence of the aether.

Einstein's solution to this was to discard the notion of an aether altogether. Light's special privilege – as something which propagates through an invisible medium and therefore must be measured with respect to that medium – was revoked. Whether you measure the laws of physics on Earth, or the moons of Jupiter, or conduct experiments in a faraway spaceship travelling at a steady speed, the results ought to be the same. Even those laws pertaining to light.

Having discarded the aether, Einstein asserted that the speed of light in a vacuum is the same for all. No matter how fast you move towards or away from a beam of light, you must always measure the same speed: 300,000 kilometres per second in a vacuum. This was one of Einstein's most famous insights and a sharp departure from the classical understanding of motion. In our everyday experience, when two objects move towards each other, their speeds add up; running towards a thrown ball, you see it coming at you faster than if you were running away from it.

How can this be reconciled? Einstein himself grappled with the problem for a long time, almost to the point of despair. After discussing it with his friend and colleague Michele Besso – the only person he thanked in the 1905 special relativity paper – he had a moment of realisation. He saw that to preserve both the principle of relativity and the constancy of the speed of light, something had to give. And that something would have to be our ancestral notions of space and time.

Speed is defined as the distance travelled within a given length of time. If the speed of light is immutable, the other two components – the distance (or space) measured and time elapsed – must be modified. The notion that space cannot contract or that time cannot pass more slowly is deeply rooted in our psyche. Yet there is no scientific basis for this feeling. As counterintuitive as it seems, as contrary to all our experiential reality, this feeling had to be dismantled to accommodate the new reality.

What the new reality demanded was a wholesale demolition of the absoluteness of space and time. Space and time were no longer immutable backdrops against which the events of the universe unfolded; there was no universal clock, keeping time for all, nor a fixed stage on which the theatre of life took place. Space and time were not independent, but relative and intertwined. In this new reality, there could be no single objective account of events, no universal 'now' that we could all agree upon; instead, the way we measured time, space, and the sequence in which events took place, was contingent upon how fast we were moving. Events that occur at the same time for one observer can occur at different times for another. Time itself could dilate, ticking more slowly for those travelling at very high speeds, while distances could contract, shrinking in the direction of motion. There was no unique, privileged perspective from which to observe the universe. The reality we perceived and measured was influenced by our relative motion.

Crucially, these effects only start to become apparent when the speeds concerned are exceedingly fast. Intuition was not going to serve us well there; we had no long-standing experience with systems moving near the speed of light and no occasion to directly observe such high velocities to integrate these effects into the realm of our common sense. The world we see and perceive at 30 kilometres an hour is not the same as that at 300,000 km/s, just as the world at scales of the sub-atomic is far removed from the scale of *Homo sapiens*. That the language of maths and science has enabled us to rise above the plane of human experience and deduce the workings of nature in circumstances so very abstract, so far removed from our own, is truly astonishing.

As is often the case with scientific discoveries, multiple minds independently converge upon similar notions, and this was especially evident in the blossoming of special relativity. Figures such as Henri Poincaré, Hendrik Lorentz and Hermann Minkowski laid the intellectual foundations from which relativity would ultimately emerge, bringing together elements that would bring us tantalisingly close to the theory – relativity seemed ripe for discovery. Yet, as the historian Olivier Darrigol remarks, while these critical components could be found in previous works, '. . . none of these authors, however, dared to reform the concepts of space and time. None of them imagined a new kinematics based on two postulates. None of them derived the Lorentz transformations on this basis. None of them fully understood the physical implications of these transformations. It all was Einstein's unique feat.'

VERIFICATION

The key implications of special relativity, the seemingly bizarre effects of time dilation and length contraction, have been validated by every experiment performed to date.

In a stunning demonstration of this phenomenon, precise atomic clocks were taken aboard commercial airliners and flown around the world twice, once eastward and once westward. These clocks measure time from the natural vibrations of caesium atoms, which are incredibly consistent and accurate. The time elapsed on these clocks was then compared to a reference clock, otherwise identical to the ones on the planes, but kept stationary at the US Naval Observatory. The results were striking: the clocks flown eastward showed a slight loss of time compared to the stationary ground clock. This is because their motion added to the Earth's rotation, increasing their speed and thus the time dilation effect. The clocks flown westward showed a slight gain of time compared to the ground clock, again because their motion was against the Earth's rotation, decreasing their speed and thus reducing the time dilation effect. It was a landmark illustration of the 'we keep our own time' phenomenon.

It's not just clocks that are affected by time dilation, but all physical processes, including biological ones. A commercial airline pilot should be slightly out of step timewise with someone with a sedentary job, but the difference would be so small as to be imperceptible. If it were possible for us to travel at some fraction of the speed of light, our bodily processes would slow down, though so too would our cognitive ability to discern this. Time dilation is manifest in all phenomena, but the speeds we typically encounter in our terrestrial existence are usually so slow that they render its effects almost negligible.

Long accustomed to a three-dimensional reality, where the three dimensions of space evolve through time, special relativity thrust us into a four-dimensional existence, one where time is an additional dimension and inseparable from space.

NOT LIKE CLOCKWORK 1

Where earlier shifts in our understanding required that we suspend our intuitions, this one also demanded that we surrender reality itself to the abstract language of mathematics, because visualising a four-dimensional spacetime is firmly *terra incognita*.

The nature of time continues to be the subject of intense philosophical debate. Is there such a thing as a 'flow of time', or is our experience of it a subjective phenomenon? Does the 'present' have any privileged status, or do all moments in time exist on an equal plane? Is the present any more real than the past and future, as our consciousness would have us believe, or is that just an illusion? Special relativity has called all of these into question.

CHAPTER 5

NOT LIKE CLOCKWORK 2

OVERTHROWING GRAVITY

After dismantling our ancestral notions of space and time and knitting them together into a four-dimensional spacetime, Einstein tried to see how to incorporate gravity into this picture. The outcome of that effort is considered to be his greatest work.

From our earliest interactions with the world, we learn that if we drop something from a height it will fall and crash to the ground. We might also intuit from our experiential reality that objects that are more massive fall faster than those that are lighter. If we were to drop a feather and a hammer at the same time, we would expect the heavier object to hit the ground first. And, indeed, it does. From this, we may reason that gravity is responsible for the difference. In fact, this turns out not to be the case. The rate at which objects fall under the influence of gravity is not in proportion to their mass; all objects fall at the same rate under gravity. In the absence of all other forces acting on the hammer and the feather, such as the air resistance that keeps the feather afloat for longer, they would both fall to the ground at the same time. This seems thoroughly counterintuitive because all our experience with everyday phenomena contradicts it, but that is mostly because we cannot isolate the effects of gravity from all the other factors influencing motion.

It was in the seventeenth century that Galileo demonstrated this proposition – which we now refer to as the Universality of Free Fall – to be true. Although legend has it that he dropped cannonballs from the leaning tower of Pisa, the experiment he actually performed may have been considerably less sensational. He rolled balls of differing masses down an inclined plane and confirmed that they fell at the same rate, regardless of their mass.

The most spectacular demonstration of gravity's universality was carried out more than 200,000 miles and several centuries away on another gravitational body: the Moon. In 1971, astronaut David Scott performed the feather versus hammer experiment during the last Apollo 15 moonwalk. In a televised test watched by millions, he dropped the hammer and feather from the same height at the same time, and, as anticipated, they both struck the lunar surface at the same instant. With no air on the Moon, there was no resistance to keep the feather from falling at the same rate as the hammer. This was reassuring, as the NASA report says, not only given the numbers watching at home, but also because the astronauts' homebound journey was crucially reliant on the validity of the theory being tested.

That gravity pulls on everything at the same rate was built into Newton's theory. But Newton's gravity also apparently acted instantaneously. If the Sun were to suddenly switch off, we would, according to Newtonian gravity, feel the effect of this instantly. However, Einstein knew that the speed of light was a cosmic limit that nothing could exceed. If it were not possible to break this limit, how could the effect of gravity be felt instantaneously regardless of the distance separating objects? Einstein reasoned that perhaps gravity isn't a force in the conventional sense at all; perhaps it isn't something that acts upon us but, rather, an intrinsic feature of the spacetime we all inhabit. He spent nearly a decade grappling with the complex mathematics needed to describe his ideas. Finally, in 1915, he arrived at a set of equations that became the cornerstone of his theory of general relativity.

General relativity contends that what we perceive as the force of gravity is actually the curvature of space and time caused by massive objects. The equations describe how matter and energy warp spacetime. Imagine how a bowling ball would create a dip if you placed it on a large trampoline, with the fabric of the trampoline as space and time. All massive objects like stars and planets are bowling balls placed on that trampoline, creating depressions in that fabric as they bend or curve the spacetime around them. This curvature then affects how other objects move. For instance, the reason Earth orbits the Sun is not that the Sun exerts some mysterious force across space, but, rather, because the Sun's massive presence distorts the spacetime around it, creating a sort of well that Earth rolls around in, similar to how some marbles would roll towards the bowling ball on the trampoline. So when objects seemingly move under the influence of gravity, they are merely following the natural curvature of spacetime. The physicist John Wheeler elegantly summarised this: matter tells space how to curve; space tells matter how to move.

As massive objects like stars and planets warp the fabric of spacetime around them, they influence the motion not only of other bodies but also the path of light. Having already discarded the idea of a fixed and absolute space and time and promoted them to an integrated spacetime, Einstein's general relativity afforded spacetime an exquisite dynamism.

Newtonian gravity had worked exceptionally well for over 200 years. Many of its predictions closely aligned with those of general relativity. It was striking that these two theories, despite being built on such different conceptual foundations, gave very similar accounts of the world. If gravity was indeed a manifestation of the curvature of spacetime, how could we put this revolutionary new idea to the test?

PLANET VULCAN

In the 1850s, Urbain Le Verrier, fresh from his triumph in predicting the existence of Neptune based on the wobbles in Uranus's orbit, turned his attention to the closest known planet to the Sun, Mercury. He noted that the orientation of Mercury's orbit, when it was closest to the Sun, was advancing more than expected by a tiny angle each century. Though seemingly small, this shift could not be accounted for by what was understood of Newtonian mechanics and the influence of other planetary bodies on Mercury. Le Verrier hypothesised that perhaps Mercury, too, was receiving a tug from an undiscovered planet that was closer to the Sun and so luminously bathed in sunlight that it remained hidden in its glare. This hypothetical planet, imagined to be even closer to the Sun than Mercury, he named Vulcan, after the Roman god of fire.

In the following decades, numerous sightings of this new planet were reported by both amateur and professional astronomers. They sought to catch a glimpse of Vulcan as it transited across the Sun's disc or during a solar eclipse, when objects near the Sun might be more easily observed. Still, the intense glare of the Sun made such observations extremely challenging; it was difficult to distinguish between a true planetary transit and sunspots or other celestial bodies. Some of the reported sightings were even contradictory. As such, there was a lack of cohesive and compelling evidence to vouch for the existence of planet Vulcan.

In 1915, as Einstein was finalising his theory of general relativity, he sought to test its validity by seeing if it could explain the observed anomaly in Mercury's orbit. After applying the complex calculations of general relativity to the problem, Einstein realised that the small but persistent discrepancy unresolved for some decades could be precisely explained by his theory. In a letter to his friend Paul Ehrenfest, a jubilant Einstein wrote, 'Imagine my joy at the result that the equations correctly yield the motion of Mercury. For a few days, I was beside myself with joyous excitement.' Indeed, this vindication of his theory gave a significant boost to its early acceptance.

BENDING OF STARLIGHT

Einstein's radical reimagining of gravity as a curvature of spacetime made specific, testable predictions that would compellingly validate its revolutionary claims. General relativity predicted that light, despite being massless, would follow the contours of warped spacetime – such that a beam of light passing a massive star would bend ever so slightly. How could we test such a prediction? We could take the largest gravitational object near us – the giant star in our cosmic backyard, the Sun – and check whether the light from distant stars is deflected by it. The Sun's brightness would ordinarily obscure such observations, so we would have to wait for a solar eclipse.

But in 1914 the world went to war and scientists who had previously worked across national boundaries suddenly found themselves on opposing sides; communication channels were disrupted, there was heightened suspicion and the flow of information so critical for the scientific enterprise was painfully stifled. German astronomer Erwin Freundlich planned an expedition to Crimea to observe an eclipse, but was promptly arrested by the Russian authorities and his equipment seized. The turbulence of the war years made it difficult to proceed, despite several opportunities for observing eclipses. Still, the dramatic new theory had caught the attention of the British astrophysicist Sir Arthur Eddington, and, after the war ended, the total solar eclipse of 29 May 1919 presented a perfect opportunity to finally put it to the test. A British expedition was organised under the direction of Astronomer Royal Frank Dyson. Eddington led a team to the island of Príncipe off the coast of West Africa, while another team travelled to Sobral in Brazil. Both locations were on the path of totality, logistically feasible and had agreeable climatic conditions, and sending two teams increased the chances that at least one group would be graced by favourable skies.

As the Moon cast a shadow on the solar disc, blotting out its light, the objective was to photograph the positions of stars in the vicinity of the Sun. According to Einstein's theory, a massive object like the Sun should cause a curvature in spacetime, which will bend the path of light from nearby stars as it passes. The apparent positions of the stars will thus shift by a specific amount, when the Sun is in the path of their light –

versus when it is not there. Newtonian theory also predicted a shift, but by a different and much smaller amount. By comparing the precise positions of the stars during the eclipse to their known positions in the sky, the teams could also verify which of the two theories the results aligned with. Where, many millennia earlier, the Babylonians had anxiously anticipated eclipses to see what fate had in store for their kings, we now eagerly awaited the eclipse to determine what it meant for how we understood gravity.

While the war had set back progress in testing the theory, it had also heightened anticipation not just in the scientific community but in the minds of the public. Drama and intrigue accompanied the expedition that set out to test relativity, and the results were announced in November 1919. The bending of starlight by the Sun's gravity was consistent with Einstein's predictions and not the Newtonian model. This landmark verification of general relativity was a major scientific breakthrough, a paradigm shift in our understanding of the universe that catapulted Einstein to international stardom. The testing and subsequent overhaul of the familiar notion of gravity captured global attention, and headlines around the world registered the decisive moment that had dethroned Newtonian physics: 'Revolution in science', 'Newtonian ideas overthrown', 'New theory of the Universe', 'Space "warped"'. Amid the post-war gloom, when the world was desperately in need of a unifying narrative, the story of a British scientist spectacularly verifying a theory proposed by a German physicist resonated as a powerful symbol of co-operation. It showed the capacity of our sense of wonder to transcend superficial divisions.

THE DEATH OF STARS

Since these early successes, general relativity has withstood rigorous testing through numerous experiments. It has made very specific predictions about challenging-to-observe phenomena that have now been empirically verified. These predictions are to do with the most violent and cataclysmic events in the cosmos, hitherto confined to theoretical constructs. Many of these cosmic phenomena pertain to the death of stars.

A star begins its life from the gas and dust accumulating in the hydrogen-rich nebulae strewn across the interstellar medium. Gravity collates this gas and dust and causes it to collapse together to form the beginnings of a would-be star. As this dust cloud collapses, it becomes hotter and hotter, and, ultimately, the core is hot and dense enough to ignite nuclear fusion, where hydrogen nuclei fuse together to form helium, and the star switches on. Its luminescence sets the cosmos ablaze with light, and radiance warms its surroundings.

Although all stars begin life in this way, the duration of their life and the manner of their death can differ greatly depending on how massive they were when formed. The more massive a star, the brighter it burns, and the brighter it burns, the faster it depletes its nuclear fuel and the swifter and more dramatic its inevitable demise. Such massive stars often die in a powerful explosion; so convulsively violent are some of these final exits that they announce themselves across the grand cosmic stage, and their brilliance and reverberations can be seen and felt from billions of light years away. In the extremities of these catastrophic events, Einstein's theory makes predictions that further distinguish it from Newtonian physics.

While a star is in the longest and most stable period of its life, as our Sun is currently, it can burn brightly for billions of years. This prolonged phase of luminosity is owed to the delicate balance between two competing forces: the nuclear fusion that unleashes tremendous energy and pushes everything outward, and the attractive force of gravity that works relentlessly to pull everything inward. This incessant duel between these forces sustains the star in a stable equilibrium. All complex life on Earth that has evolved over billions of years is indebted to the enduring stability of our own well-mannered Sun.

When all the hydrogen is used up, and there is no more to fuse, different stars bow out in different ways. Once the less massive stars have exhausted their fuel, gravity gains the upper hand and crushes and contracts the star's core. This ignites the fusion of helium nuclei to create heavier elements, and the outer layers of the star are then pushed out, causing the star to swell into a 'red giant'. The star's core doesn't collapse entirely because the electrons in the star's atoms resist the crushing pressure of gravity, and so the compression eventually stops and the star becomes a white dwarf – the deceased remnant of the star, a very dense object with, for example, half the mass of the Sun condensed into a size comparable to that of Earth, cooling slowly over billions of years. This is the eventual fate that awaits our star. In another 5 billion years, our Sun is expected to swell into a red giant, enveloping and swallowing the planets of the inner solar system out as far as, and possibly including, Earth.

Those stars that are much more massive than our Sun, however, end their days in far more violent circumstances. When they have depleted their hydrogen stores and have no more left to burn, gravity crushes their cores to a very small volume with exceptionally high density. The star switches to fusing helium for its fuel and begins to generate heavier and heavier nuclei: carbon, then oxygen, all the way up to iron. Once the core is fully iron, no more energy can be released from fusion, and the core will collapse within a matter of seconds. As it does so, it unleashes tremendous shockwaves that detonate the outer layers of the star, triggering them to blast off in a colossal explosion – the star is said to have gone supernova. This supernova explosion is so powerful and so luminescent that it can outshine several hundred million suns for many months, and the glowing remnants of the gas clouds might persist for tens of thousands of years.

It's in the crucible of these cataclysmic events that all the heavier elements – beyond iron – in the periodic table are synthesised, like gold, uranium and platinum. All of these elements, including those used to create life forms on Earth, were manufactured in the violent death throes of long-gone stars. Only in such extremities of temperatures can the heavier elements that make us, and all the structure that surrounds us, be forged.

For a star just over eight times the mass of the Sun, the end result of such a supernova explosion is a super-dense core that has around 1.4 times the solar mass packed into an area roughly 20 kilometres across. In this case, the electrons – normally found outside atomic nuclei – cannot counter the unrelenting power of gravity, which pushes them into the interior of the nucleus, where they combine with the protons to form neutrons. The process terminates in a core constituting mostly of neutrons and is called a neutron star.

But for the most massive stars, over twenty times the mass of the Sun, the core collapse is so complete that it creates an object that was theorised a long time ago but considered to be an absurdity: a black hole. In 1916, while serving in the First World War and stationed on the Eastern Front, the German physicist Karl Schwarzschild derived the first exact solutions to Einstein's field equations of general relativity. These solutions described the gravitational field surrounding a spherical mass and predicted the existence of hypothetical objects so incredibly dense that their gravity would cause an infinite warping of spacetime. This infinite curvature, known as a singularity, would form at the centre of such an object, and around it would be a critical radius within which the gravitational pull becomes so intense that nothing, not even light, can escape. This boundary marks the point of no return, the event horizon of what we now call a black hole.

While conceived as entirely abstract entities, over the decades the theoretical foundations for black holes became more and more compelling. Observations have since shown that these cosmic phenomena are not just theoretical constructs but real entities lurking throughout the universe – beginning with X-ray observations in the 1970s hinting at the presence of 'black holes' devouring matter in distant star systems. And it seems the collapse of a massive star is not the only way to create a black hole. We now think it likely that there exist black holes at the centre of most galaxies, where the dense concentration of stars are drawn together, collide, coalesce and then collapse to create a massive black hole, some of which may have consumed billions of stars. Our own Milky Way galaxy has been found to have a black hole at its centre with a mass roughly equivalent to 4 million suns.

GRAVITATIONAL MESSENGERS

Another prediction of Einstein's theory was that massive objects moving very fast would disrupt spacetime in such a way as to instigate undulating 'waves' in its fabric: waves travelling at the speed of light which would propagate in all directions away from the source. Cataclysmically violent and highly energetic events like the collisions of black holes or the explosive death of massive stars would create these ripples. The stronger the waves, the higher the likelihood that they will still be detectable by the time they reach us. Imprinted in these waves should be information about their origins and the nature of the spacetime through which they have traversed.

Although predicted in 1916, the instruments required to detect such disturbances in spacetime needed to be exquisitely sensitive and were not available until many decades later. In 1974, astronomers Russell Hulse and Joseph Taylor were using the Arecibo Radio Observatory in Puerto Rico to systematically study pulsating stars. They discovered a pulsar whose regular sweeping signal indicated that it was accompanied by another equally massive companion not very far away; the data suggested these were extremely dense stars, only tens of kilometres across but with a mass comparable with that of our Sun, orbiting each other very rapidly and closely. This was exactly the kind of system that Einstein's general relativity predicted should be emitting energy in the form of gravitational waves.

After observing this system for a number of years, Hulse and Taylor discovered that the two stars were rotating about each other faster and faster and in an increasingly tight orbit. They were able to deduce the rate of change and found it to be in remarkable agreement with that predicted by general relativity. These two densely packed stars, rapidly orbiting and closing in on each other as they lost energy, hinted at the existence of gravitational waves, but this evidence was indirect. We had still not observed the waves directly.

Then, in 2015, almost a century after the theory was conceived, the first conclusive, direct evidence for these faintest of ripples in spacetime was observed, at the Laser Interferometer Gravitational-wave Observatory (LIGO), which comprises two detectors located thousands of kilometres apart in the US (one in Washington and the other in Louisiana). More than a billion light years away, a seismic encounter had taken place that sent shockwaves through the fabric of spacetime: two black holes, one thirty-six times the mass of the Sun and the other twenty-nine times the mass of the Sun, had collided and merged. After traversing spacetime for 1.3 billion years, evidence of this colossal encounter had finally reached us, but, by now, it was exceedingly faint. To resolve it, the LIGO experiment had to be capable of detecting wobbles in spacetime that were 10,000 times smaller than the nucleus of an atom – such is the almost inconceivable precision required to detect these gravitational waves.

This monumental discovery was the first of many. In the short time since, we've detected the residual tremors from dozens of celestial encounters. Beyond resoundingly validating the bold propositions of Einstein's theory, gravitational waves have opened a new window through which we can perceive the cosmos and the dramatic occurrences unfolding many billions of light years away. With them, we've extended the instrumental toolkit through which we can 'see' the universe and paint a richer and more dynamic portrait. It would be thoroughly unsurprising if this novel tool leads us to yet more radical discoveries.

THE END OF CLASSICAL PHYSICS?

Having emphatically overturned our classical notions of space, time and gravity with a curved spacetime continuum, did Einstein's theories spell the end of Newtonian physics? Not exactly. Newtonian physics still works in many cases and can be used with remarkable accuracy for many everyday and astronomical phenomena. Where it breaks down is at the extremes. And it was only through intense scrutiny at the most extreme of scales, like the subtle bending of starlight or the slight unsteadiness in Mercury's orbit, that one framework was hailed victorious over the other. Perhaps what was most unsettling was that the new theory did not just advance the accuracy of the old but shook its foundations to the core. It was a scientific awakening that taught us how even well-established worldviews – ever so successful in explaining countless phenomena, tried and tested with utmost precision – can be unceremoniously torn down.

As Bryan Magee reflects in *Confessions of a Philosopher*, what we had to contend with was the troubling revelation that Newton's laws, despite their remarkable success, were ultimately the laws of Newton and not the laws of nature. For centuries, they had presided not just as theoretical principles that made the universe intelligible and that we lauded for their supreme elegance; we had rigorously tested them and built entire industries upon them. They had explained and predicted the motion of both celestial and terrestrial bodies with remarkable fidelity. They had accounted for the behaviour of tides in the ocean and accurately predicted planets we had no idea existed. And yet the whole edifice of our classical understanding had crumbled in the face of bold and precise new assertions that were structurally incompatible with the old order.

Newtonian physics had pointed us to the exact location in the sky to look for Neptune. We found it almost precisely where the theory said it would be. With the observed discrepancy in Mercury's orbit, we attempted the same approach, assuming that the theory was correct and that another unknown planet might be responsible for it. Yet this time, the explanation involved the wholesale replacement of the theory that had ruled for centuries.

For philosophers like Karl Popper this was a resounding affirmation of the assertion that we can never really know something to be objectively and timelessly true; that the modern lust for certainty was fundamentally flawed; that worshipping the idol of certainty was akin to worshipping at the altar of something that was illusory and logically impossible to achieve. We could not confidently assert that the laws of nature we deemed to be true were irrefutably so. We could strive to get closer to the truth, to acquire more intelligible truths, but we have no means of knowing that we've arrived at an absolute truth; we cannot claim certitude because human knowledge is ultimately human and, therefore, inherently fallible. Our conjectures could only ever be *our* laws that might closely approximate the apparent workings of nature for a defined period, and they remained corrigible as had been spectacularly demonstrated in the case of Newton.

The intellectual breakthroughs that constituted Newton's laws were not imprinted in nature and waiting for an enterprising genius to read them off; they were the construction of the human mind and a representation of how we had chosen to impose an order on the workings of nature. As such, they were radically overturned by an alternative representation that explained things better. Herein was demonstrated the power of science as a mode of thinking about the world: its enterprising ability to self-correct and remain open to falsification, as illustrated by the overthrow of Newtonian physics.

CHAPTER 6

Building Blocks 1

LAWS OF MATTER

Millennia ago, when the ancient Greeks began interrogating the world through their new rational lens, they became convinced that, despite the apparent chaos and complexity, the universe was governed by an underlying order. This order, they believed, was predicated on intelligible laws they could grasp through rational enquiry.

To understand why the world was as it was, the Greeks embarked on an intellectual journey back to the very beginning of the cosmos, seeking to decipher the fundamental substance – the raw materials – out of which all had emerged. Around the fifth century BC, the philosopher Democritus elaborated on a materialistic account of the natural world proposed by his teacher Leucippus that sought to explain the universe in terms of matter and motion, without recourse to supernatural intervention or purposeful design. They had the remarkable insight that the cosmos consisted of the nothingness or 'void' of space and infinitely many indivisible and indestructible particles they called 'atomos'. These particles differed in shape, size, order and position, and were in continuous random motion in the void, occasionally colliding and combining in different arrangements to produce the sheer complexity of phenomena we see: rocks and trees, humans and animals. But while atoms were eternal, the highly ordered arrangements they formed were temporary. Hence, the world of our experience was transitory and subject to dissolution. The conglomeration of atoms that formed us and everything around us would eventually disintegrate, and the atoms we had borrowed from the cosmos for our brief lifespans would be returned to it, where they would mill around once more until they were repurposed to form something else.

This drive to seek order amid disorder, find symmetries in complexity and uncover the unity that underlies the breathtaking diversity of the natural world is so inherent that it has underpinned all scientific enquiry since, whether it's aspiring towards a unified theory of the universe or searching for the primordial ancestor from which all life forms have emerged.

THE STRUCTURE OF MATTER

The Greeks' early attempt to describe the building blocks of a complex material world was astonishingly prescient, foreshadowing the atomic model we would arrive at much later. But this purely materialistic philosophy faced opposition from Aristotle. For him, the idea of immutable atoms contradicted the very essence of matter, which was its ability to change and transform. Aristotle had proposed that matter was composed not of atoms moving through a void but of four classical elements – earth, water, air and fire. Each element, he believed, had its natural place and movement, could change into another through natural processes and was governed not by random chance but by inherent purposes and causes, a philosophy known as teleology. The heavens, in contrast, were permeated by a fifth element, the aether – immutable and divine, unlike the corruptible and changeable earthly elements.

The Aristotelian framework, with its emphasis on order and teleology, was integrated into Christian theology through the influential works of Thomas Aquinas and remained our dominant understanding of the cosmos's fundamental substance throughout the Middle Ages.

The four-element model also deeply inspired the alchemists, who believed these elements could be transmuted and laboured tirelessly, often under a veil of secrecy, to turn base metals like lead into gold. Though their search for the philosopher's stone was elusive, the techniques they developed through tireless experimentation – grinding, mixing, distilling and heating – while primitive, foreshadowed the coming-of-age of chemistry.

In the late eighteenth century, a series of advances began to systematically dismantle the ancient belief in the four classical elements. In 1774, the British chemist Joseph Priestley successfully isolated a gas, later named oxygen, which proved to be a distinct substance, not reducible to any of the four elements of antiquity. Building on Priestley's findings, in 1789 French chemist Antoine Lavoisier, who would famously come to be known as 'the father of modern chemistry', demonstrated that water was not a fundamental element but could be decomposed into hydrogen and oxygen, substances he could not simplify any further, thereby classifying them as true elements. Lavoisier also asserted in his 'law of conservation of mass' that matter cannot be created or destroyed in a chemical reaction – it merely changes form. This principle strongly implied that there must be some fundamental unit of matter that remains constant throughout chemical processes, retaining its mass.

While Lavoisier's contributions to chemistry were nothing short of revolutionary, sweeping through the France of the time was a revolution of a different kind. Caught up in the furore of the French Revolution, Lavoisier's involvement in tax collection on behalf of the king and his aristocratic credentials made him a prime target for the wrath of the revolutionaries and saw him guillotined in 1794. Beholding this most tragic fate of a tremendous mind, mathematician Joseph Lagrange lamented that 'it took them an instant to cut off his head and a hundred years might not suffice to reproduce its like'.

These developments set the scene, in the early nineteenth century, for the English chemist John Dalton to posit what would become the modern atomic theory. Intrigued by the consistent ratios in which elements combined during chemical reactions, Dalton defined an atom as the 'ultimate particle' that constituted all matter. He visualised these atoms as billiard ball-like entities that were in incessant, random motion and indivisible – infinitesimally small but solid spheres that could not be broken down further. Every element consisted of its own unique brand of atom, identical to each other but distinct from those of other elements, and these atoms combined in fixed ratios to form compounds.

As more elements were discovered and characterised, the need arose for an organisational framework to make sense of their diverse properties.

Born in a small town in Siberia, Dmitri Mendeleev was the youngest in a family of more than a dozen children. After his father's death when Mendeleev was young, his resolute mother took great pains to provide him with an education befitting his precocious abilities. She moved the family thousands of kilometres across Russia to enrol Dmitri at a university in St Petersburg, only to succumb to tuberculosis shortly after. But the young Mendeleev remained forever indebted to his mother's sacrifices and faithfully abided by her dying words: 'Be careful of illusion; insist on work and not on words, patiently search for divine and scientific truth.' Years later, he dedicated his doctoral research to the woman who 'to give me the cause of science . . . left Siberia with me, spending thus her last resources and strength'.

One day in 1869, when Mendeleev was supposed to be inspecting a cheese factory, he used the back of the invitation letter to start arranging the known elements, over sixty of them, horizontally by their atomic weight. It seems to be then that he realised that when elements were listed as such, their properties repeated in a series of periodic intervals. Out of this work emerged the Periodic Table of Elements. This iconic visual neatly slotted into place the known elements while also leaving gaps for those as yet undiscovered, accurately forecasting their properties based on their position in the table.

When element 101 was discovered in 1955 and duly took its place in the periodic table, it was named mendelevium, in honour of Mendeleev's authoritative feat in imposing such magnificent order to the plethora of elements.

But there appeared to be no explanation as to why the periodic system had the structure it did and what underlying mechanism was responsible for arranging it thus. Shortly before the turn of the twentieth century, however, discoveries were made that resolved this mystery while shattering the belief in the indivisibility of atoms.

Physicists began prodding and probing these elemental units of matter, seeking to verify their size and structure. By observing the deflection of cathode rays in electric and magnetic fields, the British physicist Joseph John (J. J.) Thomson deduced that there existed 'bodies much smaller than atoms', with a mass of the order of a thousand times less than the hydrogen atom, that were negatively charged and appeared to be a universal constituent of all matter. To a stunned audience at a historic lecture given at the Royal Institution in London on 30 April 1897, Thomson dramatically announced the existence of this new 'corpuscle' that was far tinier than everything that had come before it. He had discovered the electron, and in doing so he cut open the solid spheres we had long presumed atoms to be and introduced us to our first subatomic particle.

Over subsequent months, Thomson furnished enough evidence in support of this claim that from that time forth 'the electron was accepted as a respectable member of society' (as noted by his contemporary James Crowther). He suggested a model of the atom as a positively charged sphere with electrons embedded in it, much like raisins in a plum pudding. This 'plum pudding' model offered an initial sketch of the atom's internal landscape.

However, it wasn't long before we had to radically redraw that portrait of the atom. In 1909, Ernest Rutherford, now considered the father of nuclear physics, conducted his famous gold foil experiment. Under his direction, students Hans Geiger and Ernest Marsden used tiny, positively charged particles (alpha particles) as projectiles, firing them at a very thin sheet of gold foil to observe how they scattered upon striking the foil. They were expecting to see almost all the high-energy projectiles – charging at roughly 15,000 kilometres per second – pass straight through undeterred.

What they discovered was astonishing: while most of the particles did indeed sail through and a small fraction were deflected, a very small number – about 1 in 20,000 – ricocheted sharply back in the direction they had come from. These were particles travelling at 5 per cent of the speed of light and the chance of this deflection happening within the existing 'plum pudding' framework was vanishingly small. It was an incredible outcome: alpha particles were known for their ability to stream through matter with little hindrance, even from the strongest electrical forces of the time. Yet, they had been made to reverse their path by a delicate sheet of gold just a few hundred atoms thick.

As Rutherford famously remarked, 'It was as though you had fired a 15-inch shell at a piece of tissue paper, and it had bounced straight back and hit you.' After months of puzzling over this unexpected behaviour, he ultimately concluded that the atom was not a diffuse cloud of positive charge, as previously thought, but, rather, had an immensely dense, positively charged core that violently repelled the alpha particles. He was also able to deduce just how close an approach the alphas were making. Extraordinarily, they came to within one millionth of a millionth of a centimetre of the centre, a distance about one ten thousandth of the atom's radius. If the atom were the size of a football stadium, the positive core was the size of a football at the centre of that stadium.

What did this mean? It meant that we had painstakingly prised open the atom to find that it was mostly, in fact almost entirely, empty space. The atom's positive charge was concentrated in an exceedingly small region, called the nucleus, which generated an intensely powerful electric field that held the electrons in place. While the atom is 99.9999999999999 per cent empty space, this great 'void' is pervaded by these fiercely strong force fields, and it is they that give matter its solidity. It is because of these forces that as you are seated, reading this, you are suspended an atom's breadth above the atoms in your chair.

With the remarkable outcome of this experiment, the plum pudding crumbled away to reveal that a miniature solar system more accurately reflected the vast distances separating the atom's constituents and the emptiness that pervaded it. But this was merely a prelude. Evidence soon emerged for even more bewildering insights into the world of the smallest raw materials of the cosmos.

The very stuff of matter, once considered the indivisible foundation of the cosmos, proved not only to be divisible but to have a rich internal structure. Within these infinitesimal entities dwelled a world far more cryptic and counterintuitive than anything we had ever encountered. Nature, in its inimitable ingenuity, had left us dumbstruck yet again. To comprehend the realm of the very small, we would have to discard all classical notions of reality, all our acquired intuitions and all desire for objective certainty. Our brain's capacity for abstraction would be pushed to the limit, and even language would show itself to be a limiting faculty in conveying the nuances of the quantum world.

QUANTUM MECHANICS

At the beginning of the twentieth century, a cascade of observed phenomena peeled away layer after layer of this secret life of matter and led us to the peculiar laws governing the smallest entities of the universe.

In the same year that Einstein published his theory of special relativity, he also proposed the idea of 'light quanta'. He advanced this as an explanation for what happens when you shine light on a metal and electrons are knocked loose from the metal atoms. It was found, puzzlingly, that electrons can be loosened if you shine blue light on a piece of potassium, but not if the light is red. Building on German physicist Max Planck's insight that energy is exchanged in discrete units, Einstein realised that light, too, was behaving as if it were made up of individual packets of energy – what we now call 'photons'. You had to deliver at least the package of energy the electron needed to break away from the atom; low-energy packets, even if delivered in abundance, would not suffice. It was this work that ultimately won him the Nobel Prize in 1921.

The phenomena we had traditionally thought of as waves could also act like discrete particles called photons. And this perplexing wave-particle duality extends to the building blocks of matter.

In 1927 the German physicist Werner Heisenberg formulated the first complete theory of quantum mechanics. One of its key tenets was that the knowledge we can obtain about a quantum system seems to be fundamentally limited by nature. There is an unbreachable limit on how precisely certain pairs of a particle's physical properties can be known simultaneously, such as its position and momentum. Known as Heisenberg's Uncertainty Principle, it states that the more precisely we try to pin down a particle's position, the more uncertain its momentum becomes, and vice versa. Crucially, this trade-off isn't due to a limitation in our tools of measurement; it is simply that there is no device, no instrument, no matter how perfect, that can breach this fundamental limit – it appears to be an iron decree of the quantum world.

Estranging us even further from our classical intuitions, quantum particles appear to have the ability to instantly influence each other, regardless of the distance separating them. This perplexing phenomenon, known as entanglement, is what

Einstein referred to as 'spooky action at a distance': when two particles are 'entangled', such that their properties become inextricably linked. Henceforth, measuring the state of one particle instantaneously determines the corresponding property of the other, and this connection seems to transcend any limitations of space and time.

In the classical picture, a particle can be said to have a definite position that we can measure. In the quantum world, things are not so simple. Rather than telling us exactly where a particle is, quantum mechanics gives us the likelihood of finding the particle in a particular place if we were to make a measurement. In other words, it provides a range of potential locations, each of them with a certain probability of being the actual spot where the particle will be found. This inherently probabilistic nature of quantum mechanics has been one of its most mystifying phenomena.

Classical probabilities arise because we're missing information about a system. Quantum probabilities are fundamentally different: they are not a reflection of our ignorance. If we were tossing a coin with complete knowledge of its initial state and all external influences on it, we could, in principle, predict with certainty whether it will land heads or tails. In the quantum realm, this is not the case. Even with perfect knowledge of a quantum system, its behaviour remains intrinsically probabilistic – that is to say, no additional information could render it predictable. Nature, at its most elemental, is 'random'.

This revelation dismantled our classical worldview, forcing us to abandon the notion of a predictable, deterministic universe and accommodate a probabilistic world where randomness is a defining characteristic. This new reality proved so alien to our sensibilities that even its chief architects,

BUILDING BLOCKS 1

including Einstein, were deeply unsettled by its implications.

So, quantum objects exist in this hazy cloud of possibilities, each with an associated probability, but what happens when we measure them? It is at this juncture that things become exceptionally murky.

Consider an electron that, prior to observation, exists as a probability cloud smeared across space. In the most widely assumed interpretation (called the 'Copenhagen interpretation'), the act of measurement pins down the electron to a definite spot from its range of potential locations. Before measuring it, we were blind to its exact state, but, in doing so, we forced it to assume a given state. In other words, quantum mechanics seems to do away with the notion of an independent observer; everything is implicated in the reality we perceive. How can the act of observation seemingly shape reality itself? This 'measurement problem' is where physics wades into the territory of philosophy, and we must confront conundrums not just about the constitution of the quantum world, but also the nature of reality and how much of it we can truly glimpse.

NATURE OF REALITY

Historically, debates about the nature of reality have given rise to two prominent and contrasting philosophical stances.

The realist perspective is that the entities described by scientific theories, even those beyond our direct senses, like electrons, exist independently of our observation. The stunning success of theories like quantum physics, with its impressive predictive and explanatory power, is given as compelling evidence for this view of the world. If the unobservable entities described by these theories were just apparitions of our theoretical imagination, the argument goes, how could we so accurately model the material world and harness its underlying principles in the technologies we have mastered? In this view, there is an external world that has properties independent of observers, and strenuous scientific enquiry can deduce those properties.

A contrasting, anti-realist perspective contends that all we ever discern is how the world appears to us. A disparity exists between the reality of nature and the mental impressions we make of it, and all we can ever perceive is this veil and not the truth of nature itself. Here, the scientific theories we have arrived at are primarily seen as useful tools for predictions and not definitive reflections of reality. Their value lies not in their truthfulness but in their utility. This view stresses the operational efficacy of scientific practices rather than the actual existence of the theoretical entities they purportedly describe.

But doesn't the spectacular success of science indicate that we are glimpsing the ultimate reality of nature? To this, one may point to the junkyard of discarded theories that litter the history of science, authoritative explanations dispensed with in favour of more plausible frameworks. For centuries, the Ptolemaic model offered a remarkably precise account of celestial mechanics, despite its erroneous geocentric premise. It was superseded by a radical worldview that shattered our centre-of-the-universe narrative. Such lessons have made us more receptive to the notion that what had once seemed incontrovertible could be replaced wholesale and overnight. Humanity has written and rewritten the story of the universe many times, and each era has, for the most part, believed the story of its time.

The philosophical debate about the nature of reality and to what extent we're able to glimpse it really comes to the fore when we're talking about nature's smallest constituents, which require us to suspend all intuitions rooted in the classical world of our human-level perceptions. Are our models of matter representative of what is actually going on or just very accurate calculational devices that we should not take too literally? Does there exist an objective reality that enterprising scientific enquiry is slowly uncovering? If, by some cosmic happenstance, we were to encounter an intelligent alien civilisation, would we be able to compare notes on, say, the electron or magnetic fields? Would their conception of these entities bear any resemblance to ours? And would mathematics prove to be the common ground for conversing about the wonders of the cosmos?

The 'measurement problem' of quantum mechanics has spawned numerous interpretations.

In the canonical 'Copenhagen interpretation' we've already encountered, a quantum system exists in a multitude of possibilities, but upon measurement it materialises in a single definite state. This is a non-deterministic and irreversible process, selecting one outcome from myriad possibilities, and it thus fundamentally accepts nature at its most elemental as probabilistic. In *The Physicist's Conception of Nature*, Heisenberg surmised that when we speak of a picture of nature, what we mean is 'a picture of our relationships with nature' and that 'science no longer confronts nature as an objective observer, but sees itself as an actor in this interplay between man and nature'.

A radically opposing alternative is to accept quantum mechanics as the literal interpretation of reality. Proposed by Hugh Everett in the 1950s is the 'many worlds interpretation', in which instead of a quantum entity taking on a definite state, every possible outcome of the measurement is realised, but in a separate, parallel world. Every time we make a measurement, there is a fork at which the universe branches into multiple unique universes, each representing a different outcome. This constant branching of reality creates the illusion of indeterminism for observers within individual branches. It is an interpretation that leaves the quantum formalism unchanged and preserves determinism, but it does so at the heavy price of grappling with an astonishingly pluralistic world.

Such divergent interpretations of the strange life of matter can be summed up in an exchange between Einstein and the Danish physicist Niels Bohr at the Solvay Conference in 1927. Bohr writes of how 'Einstein mockingly asked us whether we could really believe that the providential authorities took recourse to dice-playing, to which I replied by pointing at the great caution, already called for by ancient thinkers, in ascribing attributes to Providence in everyday language'. Werner Heisenberg, who also attended the conference, summarised this exchange as Einstein proclaiming that 'God does not play at dice', to which Bohr responded, 'it cannot be for us to tell God how he is to run the world'.

In defying our classical wisdom, quantum mechanics has forced us to concede that nature does not owe us an intuitive description of the world nor one commensurate with the shortcomings of human cognition. Throughout the history of science, we have often had to accept a counterintuitive reality. The upheavals in that history have also shown that nature has always been beyond our reach; just when we thought we'd grasped its workings, we've stumbled upon anomalies that required radical revisions to our worldview.

We don't know for how long quantum mechanics will serve as the governing reality of nature's smallest entities. Maybe future generations will supplant it with a fundamentally different narrative. Maybe they will write about our entrenchment in a worldview, characteristic of our time and era, that obscured our ability to see nature for what it really is.

In the meantime, quantum mechanics has been astoundingly successful – so much so that it has not been easy to propose viable alternatives. That would require the Herculean task of replicating a wealth of phenomena with remarkable accuracy. Our ability to manipulate matter at its smallest has practical applications that pervade our everyday, without many of us realising it. The digital age is driven by our mastery of the quantum nature of electrons; billions of tiny gatekeepers called transistors meticulously control the flow of electricity in the integrated circuits that power modern electronics. All the devices that have become seamless extensions of human existence, such as smartphones, computers and wearables, are built on the counterintuitive principles of quantum mechanics. Lasers, MRI scanners and now the potentially revolutionary paradigm of quantum computing are further examples of how our mastery of matter has, perhaps irrevocably, transformed the future of technology and the human experience.

Still, our glimpse into the world of the very small and the peculiarities of its laws is not the final word in the story of matter. Over time, it has become clear that the inner architecture of the atom – that we thought comprised the subatomic trio of electrons, protons and neutrons – may not be the full picture, but a mere trailer.

CHAPTER 7

Building Blocks 2

SUBATOMIC WORLD

In the course of developing quantum theory, the English physicist Paul Dirac realised that the maths predicted all matter to have a kind of mirror image – an antimatter equivalent. In formulating the Dirac equation in 1928, he noted that solutions to the equation hinted at particles with negative energy. This he interpreted to mean that for every particle there exists a corresponding antiparticle, which is otherwise identical to the particle but with an opposite electric charge. For the negatively charged electron, this meant the existence of a positively charged 'antielectron'. The maths was pointing us to an entire parallel realm of particles. But was the maths correct?

In 1932, Carl Anderson, a young physicist at Caltech (the California Institute of Technology, in Pasadena), was studying cosmic rays using a 'cloud chamber' – a sealed device wrapped in a magnetic field in which invisible particles leave a visible misty trail as they pass through. He took his apparatus up to the roof of the aeronautics building at Caltech; the magnet he was using required the full power of the aeronautics department's generator and had to be operated through the night because of its considerable energy consumption. While poring over photographs of the highly energetic particles that had hurtled down from space and left trails in his cloud chamber, he noticed a peculiarity. Some of the particle tracks curved in the 'wrong' or opposite direction to that expected for an electron, suggesting that a particle with positive charge had been produced. The only known particle at the time with a positive charge was the proton. But after further analysis, it became clear to Anderson that these positive particles were not the proton (which is around 2,000 times more massive) but positive electrons. Anderson had stumbled upon the electron's antimatter twin, just as Dirac's equation had predicted. This 'positive electron' was named 'positron' and it became the first antimatter particle discovered by experiment.

PARTICLE ZOO

To probe further the properties of matter and antimatter, in the 1950s scientists started building high-energy particle accelerators – machines that sent particles hurtling at near-light speeds and smashed them together. Out of these collisions and the outcomes of cosmic ray experiments emerged a line-up of new particles that we had never before encountered. The growing 'particle zoo' led to increasing speculation that, just as the periodic table had hinted at a hidden order among the elements, there may be a deeper, unifying structure governing the subatomic realm. This fuelled a fervent search for a theoretical framework to explain the observed patterns and predict the existence of as-yet-undiscovered particles.

In the 1960s, physicists Murray Gell-Mann, based at Caltech, and George Zweig, working at CERN (European Organization for Nuclear Research) in Switzerland, independently proposed that the answer lay in 'quarks' – elementary particles that combined in various configurations to form the myriad particles being observed. This quark model suggested that protons, neutrons and other 'hadrons' were not elementary but, rather, composite particles made of even smaller constituents. But after a decade of futile searches, we began to suspect that quarks may just be useful mathematical constructs without any tangible physical reality.

Were particles like the proton really structureless? Or did this nested Russian doll structure of matter have yet another layer waiting to be discovered? The answer came after a series of experiments in the late 1960s. Since Rutherford's pioneering gold foil experiment, physicists have been using particles as projectiles to infer the internal structure of matter. This time, as theoretical physicist James Bjorken put it, 'the idea was to have electrons knock protons into smithereens as violently as you could arrange it'. Instead of simply scattering off the target, the incoming projectile would impinge on the proton with such force as to break it apart. Where Rutherford's experiments could be carried out on tabletops with only a handful of others, probing deeper into the heart of matter required tremendous energy, and that meant outsized experiments and a large team.

The two-mile-long Stanford Linear Accelerator would prove to be worthy of this task. Firing high-energy electrons at protons in the Stanford Linear Accelerator showed that instead of scattering as if they were bouncing off a solid object, they appeared to be scattering as though they were interacting with smaller, point-like entities in the proton. This discovery confirmed that protons and neutrons were indeed composed of quarks. There are six types, or 'flavours', of quarks. They are called the 'up', 'down', 'charm', 'strange', 'top' and 'bottom' quarks. Each has its own distinct attributes, like mass, electric charge and 'colour' charge, that determine how they interact with the fundamental forces of nature. The lightest are the 'up' and 'down' quarks, while the heaviest by far, the 'top' quark (discovered in 1995) has nearly the mass of a gold atom. The unique properties and interactions of quarks dictate the structure and stability of atomic nuclei. Their discovery brought order to the chaotic 'particle zoo'. Starting with just these six quark types, you could explain the multitude of hadrons being observed in experiments.

Yet, nature seems to have a wildness that resists orderly classifications. Take the proton, for instance. After the discovery of quarks, we imagined the proton as a simple, manageable combination of three quarks: two 'up' quarks and one 'down' quark. But our more powerful particle colliders, like microscopes with ever-increasing magnification, meant we could pry open the inner world of the proton, which proved far more enthralling than just the jostling around of three quarks. Within the proton is a quantum sea of virtual particles and antiparticles that continuously emerge and dissipate; in addition to the familiar 'up' and 'down' quarks, there are also 'strange' quarks, and, at higher energies, even traces of the heavier quarks, all existing briefly in this quantum foam.

This complex dynamic is choreographed by gluons that interact with quarks and other gluons and, like an adhesive, bind together this hustle and bustle to keep the proton intact. The energetic exchange not only maintains the proton's integrity but also contributes significantly to its mass. In other words, this complexity isn't a smokescreen across what would otherwise be a simple structure. The proton's emergent properties, like its spin and mass, arise from the sum total of the interactions and dynamics of all its constituents. And so we need to account for not just the three quarks but also this quantum sea and the energetic gluons to explain these properties.

The inner life of protons has turned out to be so intensely vivid that people have devoted entire lifetimes to gleaning its details. That such vibrancy is apparent in even the minutest of matter is stunning affirmation that the world accessible to our senses is a fragment of a far richer reality. Our anthropocentric worldview has inclined us to believe that all the intrigue and wonder, all the complexity and colour, happens at the scale at which we exist. But nature has humbled us once more. The cosmos's building blocks have turned out to be no less wondrous than the cosmic structures they created.

UNIFYING FORCES

Throughout the history of science we've been led by a conviction that seemingly divergent phenomena may be manifestations of a single and more fundamental entity. A grand objective in physics has been to seek a unified picture of nature, where the disparate forces could be brought together under one umbrella. We know of four fundamental forces: electromagnetism, the strong nuclear force, the weak nuclear force and gravity. In the 1960s there was a major breakthrough in this quest, when two of the four forces – the electromagnetic and the weak force – were shown to be different expressions of a single interaction: the electroweak force.

In 1979, Sheldon Glashow, Steven Weinberg and Abdus Salam walked onto the stage in Stockholm to collect the Nobel

Prize in Physics for this remarkable feat. Hailing from a remote village in Punjab, Abdus Salam – who was dressed for the ceremony in traditional attire, with a white turban and gold pointy shoes – had become the first Pakistani to win a Nobel Prize.

Salam was born in 1926 into a large family with modest means, living in a two-room house with no electricity or running water. He was a child prodigy whose meteoric academic trajectory saw him win a scholarship to Cambridge University. Salam returned to Pakistan in 1951 with an earnest desire to establish a research group, but it soon became clear that he could not thrive in such an insular intellectual environment. These early experiences would make him a powerful and lifelong advocate of the need to support science in the developing world. In 1964, he founded the International Centre for Theoretical Physics (ICTP) in Trieste, north-eastern Italy, to serve as a hub for scientists from around the world and enable them to access resources and, importantly, a vibrant research community for the exchange of ideas.

As a devout Muslim, Salam saw his faith and his science as inseparable and integral to his life. Citing a tenet in Islam that encourages the contemplation of the natural world, he viewed the laws governing the universe as intimations of God's design, that 'our generation has been privileged to glimpse'. As such, Salam was despondent about the state of science in the Islamic world and often drew comparisons with the Golden Age when it had flourished so vigorously. In an impassioned bid to restore it to its former glory, he urged the oil-rich states of the Middle East and his native Pakistan to invest in the infrastructure required to realise this goal.

As the first Muslim Nobel laureate in science, Salam certainly had the credentials to lead such a revitalisation. Yet his very identity as a Muslim would become contentious. As an Ahmadi, Salam belonged to a minority sect of Islam that faced growing persecution in Pakistan. Five years before the Nobel award, the government passed an institutional amendment declaring Ahmadis as non-Muslims. Deeply troubled by these developments, Salam resigned in protest from his post as scientific adviser to the government but remained loyal to his homeland, never relinquishing his citizenship.

Salam died in Oxford in 1996, and, in accordance with his wishes, was buried in his native country. Inscribed on his tombstone were the words 'First Muslim Nobel Laureate' – but on the order of a local magistrate, the word 'Muslim' was later erased. Despite global impact and international acclaim, Salam failed to get his due in the land he loved. One might well say of him, as Will Durant once said of Spinoza: fate had written that Salam should belong to the world.

THE STANDARD MODEL

In unifying two of the fundamental forces of nature, Glashow, Salam and Weinberg laid the cornerstone for a theoretical framework that would come to dominate modern physics. Through the 1960s and 1970s the combined theoretical and experimental efforts culminated in the Standard Model of particle physics – our most successful attempt at describing the fundamental constituents of matter and the forces that govern their interactions. Like a periodic table of particles, it contains all those we consider to be the elementary (as far as we know, cannot be broken down into any smaller pieces) Lego blocks of the universe. There are twelve types of matter particles, each with its corresponding antimatter counterpart.

Of these, just three make up nearly all the visible matter that surrounds us: the 'up' quark, the 'down' quark and the electron. The rest are far more elusive, appearing only in rare and extreme conditions; when cosmic rays strike the Earth's atmosphere; at the heart of particle colliders; and in powerful natural processes like in the interior of the Sun.

One such elusive particle is the muon, a heavier cousin of the electron, that first showed up when Carl Anderson and Seth Neddermeyer were studying cosmic rays in 1936; an unexpected discovery that led the physicist I. I. Rabi to quip, 'Who ordered that?' Muons constantly rain down on Earth as a by-product of cosmic rays striking the atmosphere, and they have extraordinary penetrative power, meaning they can pass through material with little hindrance. It's this property of theirs that physicist Luis Alvarez put to use in archaeology. In the 1960s, Alvarez had discovered a plethora of new particles that expanded the 'particle zoo', for which he would receive the Nobel Prize in 1968. Meanwhile, he was fascinated by the pyramids of Egypt, particularly that of the pharaoh Khufu (the Great Pyramid) and his son Khafre in Giza. Archaeologists were mystified as to why Khufu's pyramid (with three burial chambers) has a more complex internal layout than his son's. Alvarez had the pioneering idea of using the muons streaming from the skies as powerful 'X-rays' to see if there were hidden passages or chambers in the pyramid. His team placed a muon detector in the only known chamber in Khafre's pyramid and counted the number of muons coming from all directions. If there were hollow spaces in the structure, they were expecting more muons to pass through those less dense areas. With the power of the equipment Alvarez and his team were using, they found no difference in the muons coming from any direction. There was no evidence at the time of any hidden chambers.

In a major breakthrough during 2016–17, the ScanPyramids project, using a refined version of this technique, discovered a large void, some 30 metres long – the first major internal structure found in the Great Pyramid since the nineteenth century – and a smaller corridor around 9 metres long. The abundant downpour of muons has since helped us map the internal structure of volcanoes, locate underground cavities and even find evidence of historical earthquakes in ancient burial mounds.

Beyond these matter particles, the Standard Model also includes a class of particles called 'force carriers'. These particles act much like messengers and mediate the interactions between matter particles, giving rise to three of the four forces of nature: electromagnetism, the strong nuclear force and the weak nuclear force. Gravity remains the great exception. As it stands, the Standard Model cannot explain this most familiar force.

Since its conception, the Standard Model of particle physics has stacked up an extraordinary tally of successes, explaining observations from countless experiments and predicting the outcome of many others. But it was missing a crucial member from its family of fundamental particles, and this member was critical to maintaining the integrity of this model of the universe.

In the 1960s, theoretical physicists Robert Brout, François Englert and Peter Higgs had postulated the existence of an invisible field permeating all of space. This field, they predicted, interacts with particles, and it is through this interaction that particles acquire one of their essential properties: mass. Associated with this Higgs field is an extremely rare particle, the Higgs boson – the missing puzzle piece that would validate the Standard Model and revolutionise our understanding of the universe's fundamental structure.

For decades, however, the Higgs boson remained elusive. The Standard Model did not predict just how massive the particle was and thereby how much energy was needed to produce it. Increasingly powerful particle colliders, pushing the bounds of the energy frontier, continued to see no conclusive evidence for the Higgs. That was until the most powerful particle collider humanity has ever built, the Large Hadron Collider, switched on.

PARTICLE COLLIDERS

Straddling the border of France and Switzerland, in a cavern 100 metres below the ground, is a 27-kilometre circular tunnel. This structure houses one of most complex instruments of the modern era. Popularly known as the 'Big Bang machine', the Large Hadron Collider is an engineering marvel located at the CERN laboratory in Geneva.

After the Second World War, European science was struggling and losing its top talent to the US. In 1954, CERN was established by eleven countries to foster collaboration, curb the brain drain and use science to rebuild relationships across national divides. Geneva's strategic location in Europe, Switzerland's neutrality during the war and its status as a host for other international organisations made it the ideal location for this collaborative effort. From these foundational ambitions, CERN has steadily built a reputation for world-leading particle physics, constructing a series of increasingly powerful accelerators in pursuit of matter's fundamental secrets. The latest instalment of these is the Large Hadron Collider. Designed to generate unprecedented energies, the collider effectively recreates, on a very small scale, the earliest moments of the universe. It does so by leveraging the equivalence of energy and mass that occurs at near-light speeds to create the intense conditions that existed just instances after the Big Bang.

Two beams of protons are accelerated to near-light speeds in opposite directions and then forced into a high-impact head-on collision in the interior of gigantic detectors. Like high-resolution cameras, these detectors are precisely equipped to take instantaneous snapshots of the resulting debris. The collision shatters the protons into their constituents, and these constituents then recombine and create other particles. Through studying the particles that emerge, we catch a glimpse of the raw materials that existed at the very beginning of the universe and can attempt to deduce the rules for putting them together to create the world we inhabit.

The entire apparatus that enables us to do this is a marvel of engineering, featuring a great many moving parts that each push the frontier of technology. Powerful accelerating structures lining the tunnel deliver precisely timed electric jolts to the circulating protons, gradually boosting them with every turn until they are near-light speed. Superconducting magnets chilled to an astonishing −271 degrees Celsius (colder than the depths of space) direct the particles along the trajectory of their circular racetrack. Pipes carrying the protons are maintained at an ultra-high vacuum that is emptier than outer space to minimise unwanted collisions with stray gas molecules and preserve the proton's energy.

When the Large Hadron Collider switched on in 2010, it was a monumental technological leap compared to its predecessors. With its never-before-achieved energy reach and the sheer number of collisions it was able to produce, we had crossed a critical threshold. Particle physics was now entering a realm of the subatomic world that we had never seen before. There was thus a very real possibility of either discovering the Higgs or falsifying it.

One of the first missions of the Large Hadron Collider was to determine whether our model of the universe's building blocks was going to hold sway. This required sifting through the billions of collisions that had been recorded by the detectors and identifying those rarest of collisions in which a Higgs boson might have been created. As a particle that is extremely short-lived and decays almost instantaneously to other particles, the Higgs could only be identified from the traces of what it left behind. To confirm its existence, we had to detect and measure with very high precision its distinctive trail amid the explosive sea of debris that had burst forth from the colliding protons.

By 2012, tantalising hints were emerging from the data that something new was at play. Then, on 4 July 2012, in a packed auditorium at CERN (and live-streamed across the world), two independent experiments at the Large Hadron Collider announced that a particle with properties consistent with the long-sought Higgs boson had been discovered. The room erupted; some in the audience wept. Among them was the shy and retiring Peter Higgs, now eighty-three. He had been attending a summer school in Sicily and was supposed to be en route back to Edinburgh; his travel insurance had expired and he had no Swiss francs. But he was persuaded to go to CERN for the announcement.

A year later, Peter Higgs and François Englert were awarded the Nobel Prize for predicting the existence of this crucial particle. The discovery marked a spectacular breakthrough in our understanding of nature's building blocks. It was also a major seal of approval for the theory that had for decades dominated our thoughts on the world of the subatomic. Had the experiments failed to find the Higgs boson, this would have cast serious doubt on the validity of the Standard Model, potentially falsifying it or, at the very least, necessitating radical revisions. The discovery confirmed that we are immersed in an all-pervading Higgs field. And to this invisible field we owe a tremendous debt. Its presence has enabled matter to acquire mass, an attribute without which all of the structure that characterises our material world would not have been possible.

The discovery further underscored the intensely collaborative nature of modern science. With over 12,000 scientists from more than 110 nationalities, this had been a truly global endeavour. As such, it attested not only to our universal yearning to learn nature's truth but also to the effectiveness of our collective intellect in uncovering its mysteries.

The properties of the Higgs boson are now being thoroughly scrutinised to see if it can lead us to unexplored territory. Compelling theories suggest that the Higgs could be a gateway to hidden subatomic worlds; some even posit it as the avenue to a 'dark sector' – a hitherto undiscovered troupe of particles that might explain dark matter, among other things. The elusive Higgs, once the pinnacle of a decades-long search, has now become a powerful subatomic tool to probe new phenomena that may upend our understanding of the world yet again.

ANTIMATTER

We remain mystified by the sheer dominance of matter in the universe. The cosmos, as far as we can observe, is overwhelmingly filled with matter. In the first moments after the Big Bang, we think the hot and dense universe was teeming with particle–antiparticle pairs emerging and disappearing, with an equal amount of each. But when matter meets its corresponding antimatter, they mutually annihilate. Had such a mass annihilation occurred, the universe should consist only of leftover energy from that encounter. And yet, despite our sweeping surveys of the cosmos, despite the vast distances our telescopes have scanned, in all directions all we see is matter. Why? Where is all the antimatter? From these equitable beginnings in the early universe, how did antimatter become so strikingly rare and matter so imposingly common?

Perhaps some subtle bias emerged in the early universe, a loophole in the laws of physics that allowed matter to edge out its counterpart. This bias appears to have been ever so subtle, however. For every billion particles of antimatter produced in the Big Bang, there were a billion and one particles of matter. And that extra matter particle, that ever so slight imbalance, seems to have tipped the scales in favour of matter, averting a complete annihilation that would have rendered a universe devoid of all structure. It is to that one extra matter particle out of a billion that we owe our existence, and that of stars and galaxies and the universe as far as we can observe.

So what transpired in those earliest moments after the birth of the cosmos? Where did this subtle but decisive asymmetry in matter and antimatter arise? In reconstructing those extreme conditions after the Big Bang, particle colliders also probe the processes that might hold the answer. Somewhere buried deep in nature's laws is a covenant that discriminates between particles and antiparticles. We have glimpsed this phenomenon in some rare processes; the weak nuclear force, the force facilitating the nuclear fusion powering our Sun, sometimes treats particles and antiparticles ever so slightly differently. But the observed effects are much too small to account for matter's supremacy, suggesting there is a deeper, more profound disparity we have yet to uncover.

What if there are vast swathes of antimatter somewhere in the universe, entire stars and galaxies and galaxy clusters made of it, that have so far evaded our searches? Observationally, they would look almost identical to matter; an antistar would shine in the same way a star does. But if reservoirs of antimatter exist, they are likely separated from regions of matter and, at the boundary where the two meet, there ought to unfold an intense and unmistakable cosmic phenomenon: mutual annihilation. These boundaries would be active sites of annihilation, sending highly energetic cosmic rays streaming across the universe, cosmic rays bearing the distinct signatures of such matter–antimatter collisions. As these rays strike Earth's atmosphere or spacecraft in orbit, we can study their composition to see whether they carry the telltale traces of these interactions.

Mounted on the International Space Station is one of our most sensitive instruments for studying cosmic rays and searching for antimatter; the Alpha Magnetic Spectrometer (AMS-02). Operating far beyond the protective shield of Earth's atmosphere, it has an unobstructed view of incoming cosmic rays. With powerful magnetic fields, AMS can differentiate between different types of particles, and after sifting through billions of cosmic ray events we can discern those hinting at antimatter particles. Even detecting small quantities of antimatter nuclei might suggest that large islands of antimatter exist somewhere in the universe – perhaps antistars or entire antimatter galaxies.

We have yet to obtain substantive evidence of this. Without it, there is no compelling case to suggest antimatter is present in large quantities in the universe. But our direct search for it continues, and, in parallel, we are still trying to uncover the structural asymmetry that has resulted in matter's monopoly.

Meanwhile, we've thought long and hard about the potency of antimatter and whether we can harness the energy released from its annihilation with matter. Latent in even one gram of antimatter is enough energy to rival several nuclear bombs. While antimatter has powered starships like the USS *Enterprise* in *Star Trek* and been deployed as a powerful explosive in ploys to blow up the Vatican in Dan Brown's *Angels and Demons*, in reality, it's unlikely to fuel our interstellar voyages or be a practical power source on this planet. At least, not in the near future. Antimatter is the most difficult and costly substance on Earth. Generating even minuscule amounts of it requires tremendous energy. To produce one gram demands that the largest antimatter factory on Earth, the particle accelerator at CERN, run for some billions of years, and the funds needed to do this surpass the treasuries of nations, costing many trillions of dollars. Containing it so it doesn't come into contact with matter is also highly complex. To do so, we must build highly specialised antimatter traps that keep the antimatter perfectly suspended in a vacuum, using powerful electric and magnetic fields to ensure it never touches the walls of the vessel that contains it.

The very small amounts we are able to manufacture, however, have uses that extend beyond scientific intrigue. When matter and antimatter meet, energy is liberated, and we harness this energy as an imaging technique to 'illuminate' the internal structure of the body. In positron emission tomography

(PET), a radioactive substance that emits positrons is injected into the body; when the positrons encounter electrons in the surrounding tissues, they annihilate and release energetic gamma rays. These rays are captured and analysed to build a series of slices through the body that, when combined into 3D images, reveal a vivid portrait of the internal tissues and organs. This detailed 3D visual is then used to diagnose cancer and monitor brain activity.

What was once a purely theoretical concept hinted at by the abstractness of maths has now materialised into a tangible tool for medical diagnostics. Even the most esoteric scientific endeavours can translate into practical benefits for humanity, many of them in ways we could not have foreseen, and many more we have yet to discover.

MANY-DIMENSIONAL UNIVERSE

The unprecedented conditions generated at particle colliders may also shed light on even more elemental questions, such as whether there are extra dimensions of space. While our experiential reality is confined to a three-dimensional world, there's no reason to conclude the universe is limited to this framework. In fact, according to string theory, which aims to unify the forces of nature and provide a 'theory of everything', the ultimate units of the universe are not point-like particles but, rather, vibrating one-dimensional entities called strings. In these scenarios, the mathematical coherence of the model very often requires the existence of additional spatial dimensions beyond the familiar three.

Perhaps these extra dimensions are too small and hidden from our view. And if there are many more dimensions that are not apparent to us, how can we possibly attempt to deduce them? How could we conjure up what a higher-dimensional world might look like if we're inherently spatially challenged?

In the novella *Flatland*, Edwin A. Abbott offers a charming allegory for grasping the beyond-intuitive abstraction of extra dimensions. Living in the two-dimensional world of Flatland is A. Square, whose entire existence is confined to two axes – length and width; the concept of height is notionally alien to the Flatlanders. That is until, one day, Square is visited by a Sphere from the three-dimensional Spaceland – an encounter that shatters all he knows about the reality he inhabits. If we imagine a sphere passing through a sheet of paper, it would first manifest as a point, then expand into a series of concentric circles or disks, and finally shrink back to a point before vanishing. To Square, the Sphere appears as exactly such a baffling oddity: a circle that inexplicably materialises and changes size.

Sphere attempts to explain the third dimension to Square, but, bound by the two-dimensional existence that has characterised his entire life, Square struggles to grasp the concept of anything beyond his plane. Frustrated, Sphere takes a drastic measure: lifting Square out of Flatland and into Spaceland to give him a direct, visceral experience of the third dimension. Suddenly, Square acquires an entirely new perspective; viewing his world as a plane from above, he glimpses aspects of his society that were invisible from within his two-dimensional reality.

It is similarly challenging for us to visualise dimensions beyond our own. Just as Sphere's intervention offered Square a new perspective, so we have turned to the language of maths to allow us to step outside our plane and, at least abstractly, 'perceive' these higher dimensions – an effort that has offered important insights and elegant solutions to problems that seem intractable in our familiar three-dimensional space.

Some theories contend that our entire three-dimensional universe may be like a sheet called a 'brane' floating in a much larger space called the 'bulk', which hosts not just our brane but potentially others, too. And it is through possible interactions between our brane, other branes and the bulk that we may perceive phenomena that we cannot explain through a traditional three-dimensional worldview.

For instance, some forces or particles might not be chained to our brane but be free to propagate through the bulk. Such scenarios have been posed to explain why gravity seems so staggeringly weak compared to other forces like electromagnetism; perhaps gravity is 'leaking' into these other dimensions, spreading itself thinly across the bulk, and thus appears 'diluted' in our three-dimensional world.

Other possibilities are that extra dimensions are exceedingly small or curled up so tightly that they are 'hidden' from view. Just as a walker on a tightrope is constrained to the forward and backward motion, so are we confined to our three dimensions. But if we now imagine an ant on that same tightrope, the ant can not only move back and forth but also around the circumference of the rope, which opens an additional dimension the ant has access to but that remains hidden from the tightrope walker. Similarly, some theories suggest extra dimensions that are much too small for us to perceive but that might throw up anomalies in particle colliders.

At the Large Hadron Collider, we are intently searching for the subtlest of hints that might betray that we're living in a many-dimensional universe, like an apparent violation of energy conservation from particles that appear and then disappear into hidden dimensions or the fleeting glimpses of microscopic black holes that emerge and then almost instantaneously evaporate, or extremely high energy collisions briefly 'opening up' extra dimensions and allowing exotic particles that exist there to interact with the particles in our three-dimensional world.

While seemingly outlandish, such scenarios are an intrinsic feature of theories that were proposed to explain phenomena we don't currently understand. It may very well be that nature hasn't chosen any of these scenarios as the basis of reality. If the dramatic upheavals of the last century have taught us anything, it is that nature has time and again trounced our creativity; it is quite plausible that nature has in store for us a far more 'outlandish' reality.

Meanwhile, the Large Hadron Collider and its future successors will continue to push the frontiers of human ingenuity. The coming decades will enable us to peer even deeper into the subatomic world and venture ever farther into uncharted territory. Nature's mysteries have kept us busy for many millennia, and it is evident they will do so for many more. Like an ancient relic buried under centuries of debris, our progeny might unearth the abandoned 27-kilometre tunnel that will testify to the civilisation that once dwelled here and its devotion to deciphering the ultimate building blocks of the universe. So profound has been the story of matter that Richard Feynman in his 1970 book *Lectures on Physics* mused that if some cataclysm were to befall humanity, obliterating all scientific knowledge, and we had the choice to preserve one economical but informative statement to pass to the next generation, in his opinion it would be 'that all things are made of atoms'.

CHAPTER 8

Dark Universe

If we were to journey through the cosmos we would encounter sublime beauty: dazzling worlds encircling stars, towering columns of gas and dust wherein new suns are forged, swirling galaxies of spirals and ellipses and the dark whirling vortices of black holes where gravity grants no escape. This extraordinary inventory, we once thought, must account for a significant part of the grand cosmic story. That it does not has been one of the most stunning revelations of the twentieth century.

Along with advances that have made the world more comprehensible and granted us greater control over it, the last century has also shed light on the extent of our ignorance. The most disorienting discoveries have been of the great unknowns: we've had to concede that all we've ostensibly deduced thus far accounts for a mere fraction of the cosmos. Astonishingly, a staggering 95 per cent of the constitution of the universe remains unknown to us.

All our great labour over the ages, all our striving to understand the world we inhabit, to comprehend the cosmos of which we occupy a tiny corner, has yielded insights that seem only to skim the surface. Every era has indulged in a certain hubris, believing it has achieved great leaps of knowledge, because all we can appraise ourselves against is the accumulated wisdom of the past. We now acknowledge that the matter we're made of and all the observable structures it has formed – the suns piercing our night sky, the planets sharing our Sun's warmth, the galaxies stretching far beyond our imaginations – accounts for a mere 5 per cent of the universe. The overwhelming majority remains stubbornly impervious to our understanding. Our best current guess is that over 25 per cent consists of an as-yet-unknown form of matter we term 'dark matter', and a staggering 70 per cent is pervaded by a gravity-countering force we refer to as 'dark energy'. As our most significant known unknowns, this invisible duo dominates the cosmic landscape and orchestrates the universe at large, yet they remain elusive. We recount the story of how we came to know of their existence and our theories for what they might be, while admitting that we are still very much in the dark about the nature of these phenomena.

NOT ENOUGH MASS

In the early 1930s, the Swiss astronomer Fritz Zwicky was on a fellowship at Caltech and had access to the world's largest telescope – the 100-inch. Hooker telescope at the Mount Wilson Observatory in California, which only a decade earlier Edwin Hubble had used to show that ours is not the only galaxy. Over 300 million light years away, in the constellation Coma Berenices, is a group of more than a thousand galaxies that appear to move through space like a tight-knit family – a galaxy cluster. Zwicky was intrigued to know just how gravity held together this assembly of a thousand galaxies. With the unprecedented light-gathering power of the telescope, he could peer into the depths of the Coma Cluster and the galaxies within, determine their brightness and measure their velocities. From their brightness, he could then deduce how much mass was contained in the cluster, and from the speeds of the galaxies he could ascertain the total mass the cluster must have to keep the galaxies gravitationally bound.

After calculating these figures, Zwicky noticed a stark discrepancy: the galaxies were moving rapidly, yet the mass derived from the visible luminous matter was insufficient to keep them bound together. With galaxies in the cluster hurtling through space at such incredible speeds, how had the cluster remained intact? By all accounts, this assembly should have dispersed over time; there was not enough mass, not enough gravity, to hold it together. Zwicky offered a plausible explanation. He reasoned that there must be some unseen matter that was exerting enough gravitational influence to retain this ensemble. He called this hypothetical matter 'Dunkle Materie', or dark matter – clumps of matter that did not emit light and hence were invisible to our telescopes.

Zwicky himself had a reputation for being quite an abrasive character. He was seen as irascible and arrogant and had been known to call colleagues at Caltech 'spherical bastards' – implying that they were equally disagreeable from every angle. For perhaps this reason, as well as a lack of additional observational evidence, the idea of unseen matter holding galaxies together itself lay unseen for a long time. It wasn't until four decades later, and after Zwicky's death, that the idea resurfaced in the scientific collective, this time in a rather resounding and forceful manner.

A BRIEF HISTORY OF THE UNIVERSE

VERA RUBIN

Born in 1928 in Philadelphia, Vera Rubin was an intensely curious child, captivated by the wonders of the world around her and the glorious sights above. She recalls being charmed by the light of the Moon, which loyally followed them on the drive home from her grandmother's house – 'the bushes, trees and even distant hills passed behind us, but the moon sat steadily outside my window' – prompting her to wonder, 'How could the moon know that we were going home?' While many lose their childhood wonderment, Rubin followed hers to one of the most distinguished careers in astronomy.

When she was nine, Vera's family moved to Washington DC, and the room she shared with her sister offered a clear view of the northern sky. In those days, though, the sky was exclusively reserved for the wanderings of only one gender; there were scarcely any women in the field, and while Rubin's parents were supportive, she was actively dissuaded from a career in science, advised that perhaps 'painting astronomical objects' would be a more apt profession. Rubin recalled with frustration that her high school physics teacher 'defined two kinds of discoveries: those that took insight and brilliance (here his examples all came from males) and discoveries that required hard work but not brilliance (his example was the discoveries of Marie Curie)'.

Despite the discouragement, she persevered, and after graduating on a scholarship from Vassar College, she married fellow scientist Robert Rubin and joined him for graduate studies at Cornell before completing her PhD at Georgetown University in 1954. After a decade at Georgetown and now with a young family of four children, who would all later follow their parents into scientific careers, Vera moved to the Carnegie Institute in 1965. The large observatories of the time, like Mount Wilson and Palomar, were predominantly male observatories – their living quarters often referred to as the 'Monastery', and apparently designed to be a meditative place where male astronomers could immerse themselves in the study of the cosmos without being disturbed by their families. What had initially, perhaps, been an oversight due to the lack of women in observational astronomy later became a rationale for excluding women altogether.

During the 1950s, when women applied for Carnegie Fellowships or requested telescope time, they were often denied, with housing and facilities cited as the reasons. Rubin's colleague Margaret Burbidge, best known for showing that heavier elements are synthesised in the cores of stars, managed to observe at Mount Wilson only because her husband held a Carnegie Fellowship and took her with him during his allocated telescope time. Rubin recalls the moment at the 1964 Hamburg meeting of the International Astronomical Union when Allan Sandage asked if she wanted to use the Carnegie telescope at the Palomar Observatory, which remained off-limits to women at that time. 'Of course, I said yes.' She became the first woman to legally use the Palomar telescope. Many decades later, Rubin became the first woman to have an observatory named after her and the first after Caroline Herschel – 168 years later – to be awarded the Gold Medal by the Royal Astronomical Society.

SPINNING GALAXIES

By 1968, having pushed through the social and gender barriers of the time, Rubin had published a string of papers in an area of astronomy that was becoming aggressively fast-paced and notoriously competitive. Eager to work at her own pace in a less contentious environment, she decided to turn her attention to measuring the velocities of stars in galaxies, beginning with the Andromeda galaxy, which had been studied previously but with poorer instrumentation.

Astronomer Kent Ford had devised an instrument that amplified the light gathered by telescopes such that previously inaccessible regions of galaxies could now be studied. Rubin and Ford teamed up to comprehensively survey the different zones within Andromeda and measure with unprecedented clarity how fast the stars were spinning around the centre. At just 2.5 million light years away, Andromeda is our nearest galactic neighbour and an excellent subject for such study; it looms large in the sky, spanning three degrees, so each district in the galaxy can be individually observed and a 'velocity map' of different parts of the galaxy constructed.

As Kepler had discovered and explained centuries earlier, the speeds at which planets in our solar system orbit the Sun increase with their proximity to it. The Sun's gravity dwindles with distance, and so the farther away the planet, the more slowly it needs to move to stay in orbit. Our expectation was that galaxies behaved the same way. Near the densely populated centre of the galaxy, at its gravitational heart, the stars would be moving rapidly, while those in sparsely populated suburbs would move more slowly.

What Rubin observed, however, was strikingly at odds with this. Contrary to expectations, stars residing in the outer arms of Andromeda were orbiting the centre just as rapidly as those closer to the centre. What she saw was a 'flat rotation curve', meaning their velocity remained relatively constant regardless of their distance from the galactic core. And this was not a feature exclusive to Andromeda. Rubin repeated the measurement for twenty-one galaxies that were varying distances from Earth and found the same perplexing pattern. The data were unambiguous: galaxies were not behaving how they should.

But what was the interpretation? Was this the 'unseen matter' that Zwicky had proposed as an explanation for the inexplicable Coma cluster some decades earlier? Were these swirling cities of stars embedded in a vast halo of dark matter, extending far beyond the visible galaxy and containing much more mass than that observable to us? And did the gravitational pull of this dark matter halo explain why the stars in the outer suburbs of a galaxy orbit at surprisingly high speeds?

GRAVITATIONAL LENSING

Over time, evidence kept mounting that we were missing a sizeable chunk of the universe. This became one of the biggest open questions in science, with dark matter emerging as a leading explanation. A line of support came from a phenomenon predicted by Einstein's theory of general relativity: that light bends as it passes near a massive object. And in 1937, Fritz Zwicky had proposed that galaxy clusters, with their immense mass and vast quantities of hot gas, could act as powerful gravitational lenses, bending and distorting the light from more distant background galaxies.

Imagine the ancient light from a galaxy far, far away, making its way to us across the expanse of space. If a colossal galaxy cluster were to lie between us and that distant galaxy, the cluster's powerful gravity would warp the fabric of spacetime, compelling the light from the distant galaxy to bend around it as it passes by. This bending can create striking visual phenomena: amplifying the light so the object appears magnified; distorting and stretching the light such that it appears as arcs or creates multiple images; and in the rarest of cases when there is perfect alignment, the light from the distant galaxy can be smeared into a complete ring around the lensing cluster, called an 'Einstein ring'.

Gravitational lensing is the most dramatic manifestation of the warping of spacetime by massive objects. The degree to which light bends around these cosmic lenses depends on the total mass of the object – the more massive the object, the stronger the lensing effect. In the 1980s, astronomers observed that it was far more pronounced, particularly in galaxy clusters, than could be accounted for by the visible matter alone – the stars and gas that illuminated these colossal structures. This lent further credibility to the idea that much of the mass in the universe was unaccounted for, yet its gravitational imprints were everywhere.

UNKNOWN MATTER OR GRAVITY

It was now becoming clear that an invisible scaffolding permeated the cosmos, holding together galaxies and galaxy clusters and critically orchestrating the large-scale structure of the universe. From Zwicky's pioneering observation of the Coma cluster in the 1930s to Vera Rubin's groundbreaking work on galactic rotation curves to the unexpectedly strong gravitational lensing, along with a multitude of other cosmic phenomena, a consistent portrait of the universe was emerging. This portrait indicated a significant deficiency: the total quantity of visible matter was simply not enough. In fact, it fell dramatically short of what was required to explain the observed dynamics.

If there is an unseen component whose presence we can sense only because it exerts a gravitational tug on all matter, what is its nature? We don't yet know. Multiple explanations have been proposed, but two competing classes of ideas have received the most attention. One suggests a new type of matter, fundamentally distinct from the matter we are made of – the 'dark matter' hypothesis we've been discussing. The other argues that perhaps our understanding of gravity is insufficient and in need of modification to explain these observations.

If indeed there is another type of matter, distinct from the stuff that makes us and everything around us, this 'dark matter' may be a subatomic particle or, more likely, an entire family of subatomic particles with some fundamental properties. This matter does not interact with light; its primary interaction with 'normal matter' occurs through the subtle pull of gravity, and any other interactions with our kind of matter are so incredibly feeble that they've evaded detection so far, despite our best efforts. The idea that a form of matter could outnumber ordinary matter by about five to one, making it the dominant substance in the universe, is profound – especially since we have no idea what it actually is. There could be an entire 'dark sector' that has its own structure. We only feel its gravitational tug, but what of its other forces? If the 'normal matter' that comprises 5 per cent of the universe has given rise to the breathtaking diversity of all we see, then what are the possibilities for matter that is much more numerous? To search for just one promising category of particles of 'dark matter' (called Weakly Interacting Massive Particles, or WIMPs), we've built large detectors deep underground, sent instruments into space and are even trying to create these particles in the laboratory using powerful particle accelerators.

Alternatively, the 'missing mass' puzzle may be hinting that our understanding of gravity needs revision. Since all our evidence for dark matter comes from its gravitational effects, some argue that perhaps we don't need a new form of matter but a more adequate description of gravity. Hypotheses like Modified Newtonian Dynamics (MOND) suggest that gravity behaves differently at extremely low accelerations and can potentially explain some of the observed motions of stars and galaxies without invoking dark matter. However, MOND and similar theories have so far struggled to account for the full breadth of observations.

These two ideas represent fundamentally different approaches to answering the same conundrum. How do we account for all this seemingly 'missing mass' in the universe? Where the dark matter proposition introduces a mysterious and unknown component to the cosmos, modified gravity hypotheses challenge the basis of one of our best studied and most fundamental forces.

COSMIC MICROWAVE BACKGROUND RADIATION

In the mid-1960s, on a hilltop in rural New Jersey, radio astronomers Arno Penzias and Robert Wilson were repurposing a massive six-metre horn-shaped antenna left over from the early days of satellite communications. The antenna's size and sensitivity made it suitable for radioastronomy, and Penzias and Wilson planned to use it to measure the low-level radiation emanating from gas in the Milky Way. But as soon as they set up their instrument, and before they could use it for its intended purpose, an issue emerged. They noticed substantial excess radiation that was coming uniformly from all directions and was present at all times of the day and night. It was a persistent background hiss, much like static on the radio.

They went through a catalogue of explanations to eliminate this noise or at least account for its origin. Turning their telescope towards the horizon, they scanned across the skyline, including New York City, to eliminate urban interference; they looked out at the plane of the planets to see if it was emanating from our solar system; they ruled out our Milky Way because the signal was resolutely uniform even where the Milky Way was dim. With the passing of the seasons, they could also eliminate sources that were expected to change as our planet made its way around the Sun. Even a high-altitude nuclear test conducted a year earlier, releasing charged particles into the Earth's atmosphere, should have faded significantly after a year. The pair went as far as evicting pigeons nesting in the antenna and cleaning out their white 'coating of dielectric material' – pigeon droppings.

Despite all the painstaking measures taken to identify the culprit of this signal, the radiation persisted – it amounted to an excess temperature of about three degrees above absolute zero (−270 degrees Celsius). But if this radiation wasn't from our planet or our solar system or our Milky Way, and if it wasn't an artefact of the antenna itself, then what was the source of this cosmic static? After such an extensive elimination process, Penzias and Wilson resigned themselves to the almost inescapable conclusion that this radiation came from outside our galaxy.

By the middle of the twentieth century, evidence for an expanding universe was mounting, and there followed spirited debate about its origins. In 1948, astrophysicists George Gamow, Ralph Alpher and Robert Herman suggested that the universe's early beginnings in the very distant past were smaller, denser and hotter. Such a hot and dense state would have been a perfect furnace for nuclear reactions to create the first elements that were now ubiquitous, and thermodynamics dictated that over a long period of time it eventually expanded and cooled. This relic radiation from the fiery origins of the universe would be stretched out, such that by the time it had propagated for billions of years, it would eventually reach us as microwaves. Alpher and Herman predicted that this signal should be detectable, with a temperature of several Kelvin.

Also in New Jersey, and not far from Penzias and Wilson, Robert Dicke's astrophysics group at Princeton University were building a radiometer to detect this very microwave signal predicted to be the remnants of the early universe. Penzias and Wilson were unaware of this and did not realise until they shared their findings with Dicke that they had, in effect, discovered what Dicke's group were setting out in search of. After Dicke got off that phone call, he said to his group, 'Boys, we've been scooped!'

The 'noise' that Penzias and Wilson had so arduously tried to eliminate turned out to be one of the most important scientific discoveries of all time. It was precisely the radiation predicted by the Big Bang theory. Atop that hill in New Jersey, their giant antenna had intercepted the ancient signal from the earliest murmurings of our universe some 13.8 billion years ago. This 'cosmic microwave background radiation' was the faint afterglow from the infant universe when it was just 380,000 years old.

In those earliest moments after the Big Bang, the universe

was an exceptionally hot, intensely dense and chaotic place, a seething swarm of light and matter. In this primordial soup, particles of light were continuously bumping into and scattering off a dense cloud of charged protons and electrons, unable to get very far. It was like trying to shine a flashlight in a thick fog – the light gets scattered and diffused and does not travel. As the universe expanded, it cooled, and around 380,000 years after the Big Bang it had cooled enough for the free-moving protons and electrons to combine to make the first atoms, atoms of hydrogen. With the protons and electrons now locked together in atoms, the fog cleared, and the particles of light were finally free to travel unimpeded through space. The radiation that Penzias and Wilson discovered was this 'first light', the earliest light that could travel freely in the universe, and it had been doing so for 13.8 billion years. It had stretched and cooled, and its wavelengths elongated so that by the time they reached us they fell in the microwave range of the electromagnetic spectrum.

The two teams published their findings back-to-back in the *Astrophysical Journal* in 1965, with Penzias and Wilson describing their observations and Dicke's team explaining the cosmological implications. On the front page of the *New York Times* in May 1965 appeared the headline: 'Signals imply a Big Bang Universe'. Discovering this relic radiation was indeed a resounding triumph for the Big Bang model, and edged out its rival, the Steady State model, an attractive alternative that posits an eternal and unchanging universe, with no beginning or end, where the continuous creation of matter maintains a steady density throughout.

A BRIEF HISTORY OF THE UNIVERSE

DARK ENERGY

Imprinted in this cosmic radiation is the signature of the early universe, and its detailed study can offer clues not just about that early epoch but also the parts of the universe that are currently unknown, including dark matter and dark energy. In 2009, the European Space Agency launched the Planck mission, a sophisticated observatory designed to study this radiation with unprecedented precision. Planck was sent one and a half million kilometres from Earth, to a gravitationally stable spot where it could maintain its position relative to the Sun and Earth with minimal fuel consumption. Here, away from Earth's interference, it had an unobstructed view of deep space and could make exquisitely precise and uninterrupted observations of the cosmic microwave background.

Planck produced the most precise and comprehensive map of this radiation at every point in the sky. From this, we could assemble a detailed picture of the universe from when it was only 380,000 years old. This high-resolution image pinpointed the age of the universe to 13.8 billion years. We also deduced from this radiation that the early universe was remarkably smooth and isotropic. Almost everywhere you look, you measure a temperature of very nearly 2.7K, which means that matter and energy were very evenly distributed. But zooming in on this picture shows some ever so slight fluctuations (owing to quantum effects), areas where matter was just that little bit more tightly packed. Though seemingly insignificant, these tiny irregularities had weighty implications for how the universe evolved. The areas of slightly higher density exerted a slightly stronger gravitational pull, attracting more and more matter over the course of time, gradually accumulating gas and dust and growing denser and denser. Over a long timescale, these seeds eventually grew to form the first stars and galaxies and all the structures we observe today.

Without those imperfections, the universe might have remained a vast, homogenous realm, devoid of stars, galaxies and planets, lacking all the dynamism and complexity that characterise it now. Perhaps even more astonishingly, the imperfections are fossilised into the cosmic radiation, and through measuring this radiation we have glimpsed the state of the universe when it was only 380,000 years old.

By the late twentieth century we had shored up enough support for the view that the universe had emerged from an unimaginably hot and dense plasma and has been expanding ever since, and the prevailing idea was that this expansion would ultimately be countered by the gravitational pull of all the matter within it. Our expectation was that gravity was the presiding force shaping the universe as we saw it, the chief sculptor of all the form and structure. Under its attractive influence, all matter was being drawn in, and the expansion that we were observing through the recession of galaxies would thus gradually begin to lose steam. But in the late 1990s, from studying the intensely bright exploding stars that are excellent yardsticks for measuring cosmic distances, we learned that far-off galaxies were receding faster than expected. In other words, the expansion of the universe was not slowing down, as we had presumed, but speeding up.

What could be causing this? Our mainstream hypothesis is that a strange repulsive force – 'dark energy' – is responsible for this accelerated expansion. It permeates all of space, accounting for a staggering 70 per cent of the total matter and energy budget of the universe, and effectively counteracts the attractive nature of gravity. We think dark energy may be an inherent property of spacetime itself. Even what we perceive as empty space isn't truly so. Quantum mechanics predicts that 'empty space' is teeming with activity, characterised by constant fluctuations of virtual particles spontaneously appearing out of the vacuum and instantaneously disappearing. These fluctuations have an associated energy, and, hence, so does empty space. This energy is constant throughout the universe, so it doesn't dilute as the universe expands, and we think it may be acting in opposition to gravity and exerting the pressure needed to drive the accelerated expansion of the universe. The issue is that the number our theory predicts for this energy is an astonishing 120 orders of magnitude greater than what we observe. Understanding the nature of dark energy remains one of the most elusive puzzles in physics. This mystery is all the more compelling because dark energy seems to be intimately connected to the ultimate fate of the universe.

THE FATE OF THE UNIVERSE

Within the context of our current cosmological model, we can make predictions about how events may unfold in the very distant future. The most probable scenario is that dark energy continues to drive the expansion of the universe at an accelerating rate. Over aeons, the receding galaxies will drift ever farther apart; stars will ultimately extinguish their fiery brilliance and the universe itself will devolve into a state of absolute stillness as it gradually cools towards a barren emptiness.

As this expansion progresses, in about 100 billion years large structures like galaxies and galaxy clusters will become increasingly distant from one another. The fabric of spacetime will be stretched to such an extent that individual galaxy clusters will lose sight of each other and vanish beyond one another's cosmological horizon. Surrounded by nothing but the void and unable to see anything else, each cluster will be like a standalone 'island universe'.

We're currently living through the Stelliferous Era of the universe, characterised by intense stellar activity. Stars are continually being born, shining for aeons and then dying, either in spectacular deaths like exploding supernovas or swelling into red giants and gradually fading away. Much of the energy in the universe is generated by nuclear fusion in their cores. As time progresses, the class of stars believed to play the most dominant role, which will continue to shine even when more massive stars have been obliterated, will be the red dwarf stars. These more numerous red dwarfs expend their nuclear fuel very judiciously and are expected to shine for what has been estimated to be trillions or tens of trillions of years.

Eventually, the galaxies will run out of hydrogen gas – the raw material for star creation – and star formation will cease. The last stars in the Milky Way will burn out. Even the near-immortal red dwarf stars that have been shining for trillions of years will fade away. When the universe is about 100 trillion years old, all stars will stop shining.

With all the stars switched off, the universe will be at the 'beginning of the end' phase. This Degenerate Era will be dominated by the remnants of dead stars: the cooling embers of low-mass stars like our Sun, the 'failed stars' that lacked the mass to ignite nuclear fusion, the extremely dense cores that were the terminal endpoint of supernovas, and black holes. Without any new stars being born, the skies will become dimmer, and no warmth will radiate. The universe will become a darker and colder place. Galaxies will continue to interact and merge, potentially yielding supermassive black holes at their centre. Some theories suggest that the subatomic particles, protons, will eventually decay, and in this case even dead stars will eventually disintegrate and disappear. At the end of this era, all that will remain are black holes, and these will come to dominate as the Degenerate Era gives way to the Black Hole Era. Since they are unaffected by proton decay, black holes will be the only star-like objects to survive the preceding era and the universe's primary source of power. But over a long enough timescale, even black holes will evaporate through a process called Hawking radiation, which will see them converting their mass into radiation.

This will plunge the universe into an epoch called the Dark Era, characterised by sheer desolation – a colossal vastness, an all-encompassing coldness and utter emptiness. This scenario could unfold around 10^{100} years after the Big Bang – that's one followed by a staggering 100 zeroes – a timescale beyond the grasp of our comprehension. Only scattered particles remain, separated by immense distances. All energy is evenly distributed, and matter and energy are in such a diffuse state that no further work or interaction can transpire. Devoid of all form and structure, the universe will succumb to eternal nothingness, a void. This is called the Big Freeze, or the Heat Death of the universe.

There also exists an alternate possibility in which the expansion of the universe doesn't just continue but actually accelerates with time. Some models propose that if the strength of dark energy increases over cosmic epochs, it could become so formidable that it overpowers all other forces, including gravity. Ultimately, this would mean that galaxies are torn apart, stars are violently ejected from their stellar families, planets disintegrate and even atoms are shredded into their most elementary constituents. This cataclysmic vision has been termed the 'Big Rip', an end where all cosmic structures are irrevocably dismantled into their fundamental components.

The heavens have changed little since the ancients began charting them, but humanity's perception of them has altered a great deal. Where the Aristotelian worldview established a clear distinction between the death and decay on Earth and the eternal perfection above, we now know that stars also grow cold and die. The universe itself may be heading towards a similar fate. While we live under the same sky, we inhabit an altogether different reality. And our descendants will occupy a different one still.

By now, we're more attuned to the idea that the 'truths' we hold to be indisputable are liable to be swept away in the next scientific revolution. Arthur Eddington once likened scientific discoveries to putting together the pieces of a great jigsaw puzzle. A new discovery may not mean that the already arranged pieces have to be taken apart but that the full picture may need to be reinterpreted. 'One day you ask the scientist how he is getting on; he replies, "Finely. I have very nearly finished this piece of blue sky." Another day you ask how the sky is progressing and are told, "I have added a lot more, but it was sea, not sky; there's a boat floating on the top of it." Perhaps next time it will have turned out to be a parasol upside down.'

Our Search for Cosmic Company

CHAPTER 9

Life As We Know It

For centuries we've looked up at the silent starriness of the night sky and wondered, are we alone in our celestial voyage? Or are there alien beings gazing at their night skies asking the same question? We've long clung to the idea that our existence holds a special significance in the cosmos – a belief that has slowly unravelled as the universe unfolded before us. We now find ourselves circling an unremarkable star in the spiral suburbs of a sprawling galaxy, one among 2 trillion in an expanding universe. By now, any Earth-centric notions of exceptionalism have been thoroughly dismantled; our place in the cosmos appears no more special than that of a single grain of sand randomly scattered on a beach.

But what about life in the cosmos? How unique or universal is it? Where exactly do we reside in the biological scheme of the universe? Of all the dramatic upheavals that have unseated us, discovering life beyond our planet would be sensationally transformative, forever altering how we see ourselves.

Millennia ago, the Greeks pondered the plurality of worlds. While the Aristotelian camp held that Earth was unique and admitted no other worlds besides it, the atomists adopted a pluralistic stance and argued that there must exist innumerable realms. They reasoned that if worlds were formed from the accretion of atoms, it logically followed that there were others where atoms had similarly come together. Democritus and Leucippus were leading proponents of this view, which was later embraced by the Epicureans, who believed in an infinite universe consisting of countless indivisible atoms and an endless expanse of empty space through which these atoms move.

It was the Roman poet and philosopher Lucretius, in the first century, who gave the Greeks' atomic theory a thorough exposition in his celebrated treatise *On the Nature of Things*. In a world governed by mechanistic laws rather than by gods, and consisting of only atoms and the void, he reasoned that: '. . . when there is such an ample supply of matter, so great a store of elements, and no stint of space, it is not likely that only this one earth and sky has been created, and that those bodies are doing nothing elsewhere . . . there must be other assemblages of matter similar to this one of ours, which the ether holds in its embrace.'

Throughout history, people have speculated that this planet of ours may not be and perhaps cannot be the only one, but the influential Aristotelian framework exalted us as unique and at the centre of all there was, and this became the prevailing worldview for millennia. It took the exiling of Earth from the centre of the universe for notions of the plurality of worlds to gain traction. In the sixteenth century, the Dominican friar Giordano Bruno speculated that the stars in the firmament were, in fact, distant suns, and that there were innumerable earths, each circling their own sun and possibly harbouring their own life forms. He also held dissenting opinions on the core doctrines of Christianity that were deemed heretical by the Church. After seven years of imprisonment and interrogation, he was found guilty of heresy by the Roman Inquisition and executed in 1600. It was only after his death that he gained widespread popularity and came to be seen as a martyr for free thought.

In the centuries that followed, the idea that there may be worlds like ours out there in the vastness became much more plausible. Only relatively recently have we acquired the means to catch a glimpse of them and earnestly probe the question of whether we're alone, or if there exist others like us.

However, before we embark on a search for life beyond Earth, we must first grapple with a fundamental question: what exactly is life, and is our definition of it broad enough to encompass the diverse forms it might have taken elsewhere in the cosmos? And how did life as we know it arise on our planet and acquire the capacity to question its own existence and the nature of the cosmos it inhabits?

WHAT IS LIFE?

At one time, the only explanation for the enigma of life was that supernatural intervention had graced it with a unique essence, giving it distinction in a lifeless world. Life was spirited, where inanimate matter seemed sterile; life moved and stretched to catch the rays of the blazing sun, while the non-living stood inert; life appeared restless, seeking sustenance to grow and make copies of itself; and above all, life was vulnerable – the vitality and dynamism that characterised it eventually disintegrating, rendering it indistinguishable from the inanimate matter from which it had arisen.

For many millennia we have pondered the marvels of existence and their emergence from seemingly ordinary raw materials. Yet, a satisfactory definition has eluded us. Is life fundamentally a biological phenomenon? Or is it ultimately rooted in the immutable laws of physics? Or perhaps it is better viewed through the lens of information processing? Some thinkers have asked if the phenomenon we label life is not so distinct from non-life, that these binary polarities we have imposed are misleading and simplistic, and the living and non-living may exist on a spectrum that we have yet to fully understand. Perhaps the spark of life is not in fact a sudden event but, rather, a gradual intensification along this continuum. With the recent emergence of artificial forms of intelligence and their possible future, it is evident that we will soon be forced to delineate between life and non-life, sentient and non-sentient beings. We may only be a generation or two away from thinking of life in a very different way.

The widespread definition adopted by NASA in the search for life beyond Earth is that 'life is a self-sustained chemical system capable of undergoing Darwinian evolution'. It asserts that life is fundamentally based on chemistry, that it has the functionality to sustain itself, and that within this self-sustaining framework it admits some random variation that can be acted on by a mechanism like natural selection to maximise an organism's ability to survive and reproduce in a particular environment. This working definition aims to be expansive enough to permit the possibility of life forms that may have followed a different blueprint from that on Earth. But there are over a hundred other definitions of life. This multitude of interpretations shows the inherent challenge in defining something for which we have only a single template.

ORDER FROM DISORDER

In 1944 the Austrian physicist Erwin Schrödinger, best known for his role in developing quantum theory, turned his attention to the enigma of life. In his influential book, *What Is Life?*, Schrödinger brought his profound insights to bear on a question that had primarily resided outside the domain of physics.

What most fascinated him was how life *appears* to break one of the most inviolable of nature's laws: the second law of thermodynamics. This law dictates that the entropy, or disorder, of a closed system inevitably increases over time – that there is a natural tendency of things to advance towards a chaotic state. To this law we owe that heat flows naturally from a hotter object to a colder one and not the other way around, that a neat and tidy room will become messy and cluttered over time and not the reverse, that the universe itself is on a one-way journey towards a state of maximum entropy, the ultimate cosmic heat death where all energy becomes evenly dispersed and all activity grinds to a halt.

So supremely held is this law that Sir Arthur Eddington once declared that if your cherished pet theory contradicts other well-established principles of physics, there may be room for reconciliation, but 'if your theory is found to be against the second law of thermodynamics I can give you no hope; there is nothing for it but to collapse in deepest humiliation'.

Life prevails as a marvellously intricate and well-ordered system, that, at first glance, seems to defy the universal trend towards disorder. If the second law is an inviolable decree of nature, how does life seemingly and so boldly violate it? The answer is that it does not, because the second law applies specifically to isolated systems, those that don't exchange matter or energy with their surroundings. While the universe as a whole is deemed an isolated system, our planet and the living entities that reside on it are not – they are constantly making energy transactions with the Sun. The second law permits local pockets of order, as long as the overall disorder in the universe is increasing. It is this loophole that life exploits.

Schrödinger's insight was to recognise that living organisms, with their remarkable complexity, maintain their internal structure by continuously extracting 'order' from their environments. In essence, life makes the following bargain with the universe: in exchange for the orderliness that allows it to exist, it will spread disorder. Living things take in energy and matter from their surroundings in a more ordered state (like sunlight or food) and use it to fuel their own processes. The by-products of these processes, such as heat and waste products, are released back into the environment in a less ordered state, thus increasing the overall disorder of the universe. It's a transactional affair and a dynamic one, requiring a perpetual exchange of energy to keep the lights of life on. Persistent bartering with the universe to maintain our orderly conglomeration of atoms and stave off inevitable decay is what characterises life from this perspective.

To fulfil this bargain, plants capture the energy of sunlight and convert it into chemical energy through photosynthesis. Animals then consume the complex, organised molecules in plants or other animals and break them down to extract energy for their own bodies, releasing waste. In effect, animals 'borrow' the order created by plants and dissipate it into the environment as disorder. Living entities thus cleverly evade the universal tendency to disorder by engaging in a frenetic exchange of energy. It's a dynamic yet delicate affair and not one that can be sustained for long.

In 2013 the physicist Jeremy England extended this perspective with a provocative idea: what if life is actually an expected outcome of the universe's fundamental laws? He posits that if a collection of atoms were basking in the warmth of the sun, they would spontaneously arrange themselves in ways that are increasingly efficient at capturing and channelling that energy – because matter, when exposed to energy, has this tendency to self-organise into intricate assemblies that excel at dispersing that energy. Over time, this process would yield the increasingly complex structures that we recognise to be living things. In other words, a plant exists because it is far more efficient at dispersing energy than a random heap of carbon atoms.

Living organisms, then, are like 'disorder-creating machines'. If we accept the cardinal rule of the universe – that the overall net disorder must increase – then it isn't unexpected that life should arise. Even life's defining characteristic of making copies of itself can be seen as a strategy to enhance energy dissipation, a way to create more 'disorder-creating machines' and multiply its ability to disperse energy throughout the environment. Rather than being an anomaly or an accident, life becomes an expected consequence, an emergent property of the laws of physics and not a distinct phenomenon apart from them. Life, in this framework, is not just compatible with the second law of thermodynamics, it's essentially driven by it.

While the theory is still being explored and debated, it offers a compelling alternative for understanding the essence and emergence of life, which may have implications for how prevalent it is in the universe. If life is a natural consequence of thermodynamics, then it might be far more common and widespread than we previously thought.

LIFE IS INFORMATION

An even more abstract perspective of life is to view it through the lens of information, recognising that what really sets life apart is its enterprising use of information to sustain itself and adapt to changing conditions. In this framework, life is analogous to a processing unit that can store, organise and replicate information. Rather than being a biochemical machine that is running a genetic programme, life *is* the programme and the biochemical structure is just the way it is physically manifested.

Living beings are voracious consumers of information; they assimilate and integrate data both internally from within their system and externally from their environments. It is through this dynamic flow of information that they adapt to their changing world. In this model, every component of a living system plays some role in processing and using information; cells talk to each other, organisms interact with their environment and entire populations adapt and evolve over time in response to external events. At the molecular level, information is encoded in the molecules of DNA or RNA, which carry the blueprint for making proteins. This information is given meaning through the context in which it is employed; a DNA molecule has no function on its own, but within the context of a cell it contains the instructions necessary to build proteins essential for life. At the cellular level, information is processed within the cells themselves, co-ordinating various activities and allowing them to function. At the level of the organism, a larger network of information is set up so cells in different parts of the body can communicate with each other to maintain the integrity of the organism and allow it to grow and develop. At the ecological level, information flows between the organism and its environment, shaping the ecosystem and the organism's ability to adapt. Ultimately, life perpetuates itself by transmitting this information to the next generation through reproduction. Self-replication can be seen as a mechanism for taking rare and valuable information and making it more abundant.

This framework for categorising life may become increasingly prominent in the technological age as we think about intelligent systems that do not require a biological substrate.

LIFE AS WE DON'T KNOW IT

If we cannot even define life and have only a single blueprint for it, how can we possibly contemplate the forms it might have taken on other worlds? This remains a formidable challenge. A well-known maxim asserts that the first place to look for lost keys is under the lamppost. Our notions of, and search for, extraterrestrial life have taken a similar approach.

We owe our existence to the scattering of the ashes of dead stars, which spewed heavy elements into the cosmos from which the complex chemicals that characterise life became possible. It took the demise of several generations of stars to acquire the raw materials to produce us. Owing to this stellar alchemy, the building blocks for life as we know it – carbon, hydrogen, nitrogen, oxygen, phosphorus and sulphur – are abundant in the universe. We are famously 'made of starstuff' and, since starstuff is in plentiful supply, it's not altogether unreasonable to wager that life elsewhere might have drawn from similar ingredients. Although we are a product of the universe, we are also fundamentally a product of our planet and the specific conditions, past and present, that have forged our unique biochemistry. Unable to extricate the two, we have to question which of the ingredients for life are truly universal.

For instance, does life require water? Water's unique properties have made it critical on Earth. It's an excellent solvent for transporting nutrients and waste, remarkably stable and remains in liquid form over a wide range of temperatures. Yet it's possible that solvents like ammonia or methane can also support chemical reactions, though they would give rise to a biochemistry unlike anything we're familiar with. Similarly, does life have to be carbon-based? This fourth-most abundant substance in the universe is also chemically incredibly versatile, forming stable bonds that have made it the element of choice for living things on Earth. But there is now increasing interest in the parallel chemistries of elements like silicon that have at least some of carbon's attributes and may have been adopted in other planetary environments. The same goes for storing genetic information: Earth's organisms use DNA and RNA but this could be just one chemical basis among many that are capable of passing information from one generation to the next.

While our conceptions of extraterrestrials are fundamentally biased by the only life we know, we are beginning to challenge those assumptions and ask: How different could life be?

Now, though, we turn to the singular template we have of life in the universe and chart the path it took from its simple unicellular existence to a space civilisation. This reconstruction of life on Earth takes us through a labyrinth of improbable and apparently unforeseeable twists. Our existence on Earth was contingent on a staggering cascade of improbabilities. There was certainly no inevitability about it.

LIFE ON OUR PLANET

The story of life on our planet goes back many billions of years, and its origins are obscure. We still don't have a definitive consensus on how life came to be, though multiple narratives abound. The process by which non-living matter transitioned into the first self-replicating entities remains a most mysterious phenomenon.

Did life arise near hydrothermal vents deep within the ocean, where hot, mineral-rich fluids interact with cold seawater to create a unique chemistry for forming complex organic molecules? Did it originate in a primordial soup in shallow pools or oceans, where lightning strikes or ultraviolet radiation from the Sun triggered the chemical reactions to generate life's building blocks? Or did life originate elsewhere in the universe and hitch a ride to Earth on comets or meteorites? This panspermia hypothesis doesn't explain the origin of life, only that it arrived on our planet from somewhere else, thus shifting its origin to another place in the cosmos.

We cannot as yet rule out the possibility that there were multiple, independent origins of life on our planet, but if life did arise on other occasions, we have yet to come across any of its descendants. It is plausible, however, that long before the first fossils appear in the record, life might have tentatively found its footing, only to be swept away by the chaos of early Earth, where volatile conditions would have made it exceptionally challenging for life to retain its foothold.

A BRIEF HISTORY OF THE UNIVERSE

Not long after our planet formed, around 4 billion years ago, we think the universal ancestor of all living organisms we know today emerged – termed the Last Universal Common Ancestor, or LUCA. That all known life forms came from this single lineage seems strikingly presumptuous, but we have compelling evidence to support it. For one, all organisms on Earth carry the same underlying blueprint – a shared genetic code that provides the basic instructions for building life in all its diverse forms. Secondly, many fundamental biochemical processes, such as the way we break down sugars for energy, exhibit remarkable similarities across all domains of life. Thirdly, an almost universal set of twenty amino acids is used, and although they exist in both their left- and right-handed mirror images, life has chosen only the left-handed form. And, perhaps most fascinatingly, the hierarchical pattern of relatedness between different species, represented in 'the Tree of Life' (an organising framework that depicts the relatedness of different organisms and their evolutionary history), shows the interconnectedness of all life, pointing resolutely towards a single common ancestor at its root. It's astonishing that life on this blue sphere, from the smallest microbe to the gargantuan dinosaurs, can be traced back to this origin. Probably highly resilient and capable of withstanding extreme environments, the emergence of LUCA set in motion a cascade of breathtakingly improbable events that would eventually lead to us, *Homo sapiens*.

MAJOR LEAPS

The first such event occurred around 2 billion years ago, with the birth of eukaryotes, the building blocks of all complex life on Earth. In the microscopic world of single-celled organisms, which had ruled the Earth's biosphere for almost 2 billion years, a chance encounter took place between two prokaryotic cells (cells lacking compartmentalised structures within their cell membranes), where one of them, probably the larger of the two, engulfed the smaller. Rather than being consumed and digested, the engulfed cell found itself forever trapped inside its host and eventually, an unlikely symbiotic partnership developed between the two. The enslaved cell became the energy powerhouse – the mitochondria – supplying its host with a steady stream of energy, while the host cell provided a sheltered environment for its new ally. This alliance forged the foundation for the eukaryotic cell.

For over a billion years after this, our planet's appearance altered little; life was still resolutely microscopic, and single-celled creatures remained uncontested in Earth's biosphere. Then, after another billion years, life took another unprecedented leap in complexity, as single cells joined forces to form complex, co-operative assemblies. This marked the evolution of multicellularity, when individual cells came together and specialised to perform specific functions. Cells were differentiated into distinct types: some became proficient at harvesting energy from the Sun, some at breaking down nutrients, and others took up the task of carrying oxygen or fighting off infections. This ability to specialise led to the emergence of complex systems, like tissues and organs, that comprise larger organisms like us.

Multicellularity independently evolved around twenty-five times in total. Although the particulars remain a mystery, it is evident that once the threshold of multicellularity was crossed, life's narrative on Earth grew richer and more varied, spawning life forms with an extraordinary range of sizes, shapes and ways of functioning. Natural selection crafted an array so diverse and wondrous that the event is called the Cambrian explosion. In this dazzling biological renaissance around 550 million years ago, we see a sudden burst of evolutionary creativity as most of the major animal groups start to appear in the fossil record. Unlike the earlier soft-bodied creatures that left behind minimal traces, these animals had well-preserved hard parts, like shells and exoskeletons, making the Cambrian explosion seem even more dramatic.

Not only do many of the major animal groups appear during this time, but the first instances of body plans and adaptations, like the first animals with eyes, legs and antennae, also start to appear. Evolution now had enough tools and tricks to craft increasingly complex life forms. In the wake of the Cambrian explosion, the first ancestors of vertebrates appeared, giving rise to jawed fish and then to the pioneering tetrapods. Although initially adapted for aquatic life, these four-limbed creatures eventually ventured onto land – one of the most consequential migrations in the history of life on Earth.

These early land-dwellers gave birth to lineages of amphibians, reptiles and mammals. Though mammals emerged around 200 million years ago, they remained relatively small and inconspicuous, lurking timidly in the shadows through the era of gigantism. That was, until around 66 million years ago, when a ten-kilometre-wide asteroid hurtling towards Earth at 45,000 miles an hour struck the Yucatán Peninsula in Mexico and, it appears, triggered a mass extinction that annihilated three-quarters of plant and animal species, including non-avian dinosaurs that had reigned for over 165 million years. This cataclysmic event forced a large-scale reboot of life on Earth. With so many long-held ecological niches now vacant, a remarkable diversity of new forms and species rushed to fill them. It was in this vacuum that mammalian life began to thrive.

THE EMERGENCE OF SKY GAZERS

Around 6–8 million years ago, the branch that ultimately led to modern humans diverged from that of the great apes, specifically the chimpanzees. The closest of our living relatives, with whom we share a remarkable 99 per cent of our DNA, are chimpanzees and bonobos; in fact, we are more closely related to them than they are to gorillas. But over those 7 million years we gradually acquired a catalogue of physical and behavioural traits that moulded us into our human form. From walking and running on all fours, we adopted an upright, bipedal gait, freeing our hands to carry goods, craft tools and signal and gesture to one another. With this newfound dexterity, we migrated from dense jungle environments to the open savannahs, where standing tall granted greater visibility and the agility to traverse large distances more efficiently. Having spent aeons with our gaze fixed on the Earth, we could also now turn our face to the heavens, where the shimmering lights that cut through the darkness would become our preoccupation.

Over the past few million years, various hominin species emerged, some with a fusion of ape and human traits, such as ape-like arms suited for gravity-defying acrobatics among trees coupled with brains demonstrably larger than their predecessors. The birth of the *Homo genus* around 2.8 million years ago marked a defining chapter in our evolutionary story, a story that is messy and complicated. We're still trying to piece together the various branches of our family tree, figuring out the multitude of lineages that existed, the cul-de-sacs that were evolutionary dead ends, and what the precise relationships were between human species.

Among the earliest of our relatives to have an intriguingly human-like face and body was *Homo erectus*, or 'upright human'. If endurance is a hallmark of evolutionary success, *Homo erectus* was extraordinarily successful – fossil evidence for their existence stretches back more than 1.5 million years, making them the longest surviving of all our human relatives. They are also considered among the first known hominins to set forth from Africa and, possibly, the first to control and use fire.

Over time, our ancestors' cranial capacity generally continued to expand, and their mastery over stone tools and technology grew ever more sophisticated. Living roughly between 700,000 and 200,000 years ago, *Homo heidelbergensis* were communal hunters and built rudimentary shelters, early manifestations of the complex social behaviour that would ultimately define our species. As *Homo heidelbergensis* continued to evolve, over time they branched into other lineages. In Europe, one of these lineages probably gave rise to the Neanderthals, a species better adapted to the harsh, cold climate of the Ice Age. Meanwhile, in Africa, another lineage would ultimately give rise to *Homo sapiens*.

It was around 300,000 years ago that early modern humans emerged in Africa, bearing a striking resemblance to us in many key aspects and possessing a suite of unique traits that set them apart from their predecessors. Over time, they began to craft increasingly complex tools from stone, bone and wood; express their creative flair and capacity for abstract thought in cave paintings and carved figurines; come together in co-ordinated units to hunt larger game; probably developed divisions of labour; and transmitted cultural wisdom from one generation to the next. Such advancements laid the foundations of intricate social structures, paving the way for the eventual emergence of the civilisations that have been so definitive in the trajectory of humanity.

Although we are now the sole representatives of our kind, at least nine human species once thrived on this planet, including the Denisovans, who lived in what is now Siberia and Tibet, and *Homo floresiensis* (nicknamed 'the hobbit' because of their short stature), who inhabited the island of Flores in Indonesia. When our ancestors migrated from Africa and spread across the globe, it appears they encountered other lineages, such as the Neanderthals and Denisovans. So similar were these cousins of ours that we even mated with them and to this day, carry evidence of those encounters in our genetic heritage. Modern Eurasians can trace around 2 per cent of their ancestry to Neanderthals, while Oceanians have around 2–4 per cent Denisovan DNA. Intriguingly, the genome of modern West Africans also contains the legacy of a hitherto unknown species, hinting at a more diverse family tree than we currently understand.

For about 40,000 years, however, there has just been *Homo sapiens*. All the other human species eventually lost their foothold and became extinct. We don't know for sure what happened to them. Perhaps a constellation of factors conspired to bring about their eventual demise: climate change may have fragmented their groups and prevented them from establishing large population bases; their smaller population sizes possibly led to excessive interbreeding, making them more susceptible to disease and also less socially resilient; and perhaps they were unable to outcompete the slightly more capable newcomers – us.

As uncontested leaders of the primate world, our exclusivity as the last of the human species has indubitably coloured our perspective. If our human relatives were still around today, following parallel trajectories and possibly exploring the cosmos as we do, our sense of self-importance might be greatly diminished. Even subtle alterations in our evolutionary story could have drastically changed the outcome: perhaps there would be no musings on our place in the cosmos, or our Neanderthal cousins would have penned them instead.

From the narrative we have pieced together of how life evolved on Earth, it is evident that although simple life emerged soon after conditions became conducive, the route to complex, conscious, cosmos-enquiring life forms happened against staggering odds. A series of highly improbable and unpredictable events had to stack up to create the perfect crucible to forge *Homo sapiens*. Natural selection had to incrementally sift through a great many species and labour over billions of years to craft an entity capable of self-enquiry. There was certainly no inevitability that once the threshold for life was crossed, a space civilisation was a foregone conclusion. We were not the end goals of evolution, because evolution has no goals. Of the billions of creatures Earth's biosphere has cradled, only one has acquired the means to venture beyond it; it has been a statistically costly exercise, requiring a large denominator of life forms to produce but one example. And, even then, our survival was not guaranteed. Our relatively low genetic diversity, compared to even a small band of wild apes, suggests that our species went through a near-catastrophic crash at some point in our evolutionary history during which our numbers plummeted to dangerously low levels. In other words, we very nearly did not make it.

How many worlds are there where life opened its eyes but did not reach the stage of acquiring sophisticated language and culture to immortalise its existence or to venture into the cosmos in search of others like it? Yet there was so little separating us from others in the animal kingdom. We can only speculate whether intelligent life elsewhere in the cosmos had to negotiate similarly impossible odds.

To imagine what life might look like beyond Earth, we must liberate our imaginations from the familiarities of our terrestrial existence, which is not altogether easy. If we had only ever encountered those of our own kind, would we have been able to envisage the breathtaking diversity of forms that nature has crafted here? Would we have imagined creatures capable of flight, dwellers of the deep sea, the intricacies of ant colonies, the world of the microscopic, or the gigantism of dinosaurs? Thanks to the vast multiplicity of life on Earth, our horizons for life have been greatly expanded, but we may still be inherently limited by the specific biology that has taken root here.

OTHER WORLDLY LIFE

To step outside these confines, we have looked to entities on our planet that are profoundly 'otherworldly'. One subject of interest has been the octopus. The cephalopod family – octopuses, squid and cuttlefish – have followed a divergent, almost independent trajectory from our vertebrate lineage. Our evolutionary branch separated from theirs nearly 750 million years ago – our last common ancestor was a primitive flatworm that trawled the sea floor. Therefore the octopus, along with the others in the cephalopod family, embodies an intelligence that took a very different path from our own. It is often described as the closest we can get to an alien intelligence – our 'terrestrial alien' – and hence an excellent muse for considering the possibilities of intelligence beyond our planet.

After they lost their protective shell, octopuses evolved a range of mechanisms to evade predators. Their adaptive skin can change colour and texture to masterfully mimic their surroundings – be it rocks, seaweed or the sandy seabed – and, when threatened, they unleash a cloud of ink as a smokescreen to disorient predators and allow for a swift escape. For quick manoeuvring, they are equipped with a jet propulsion system that can expel water through a muscular tube, enabling them to change direction with incredible agility and even propel themselves short distances out of the water.

But while their body plan is distinctly different from our own, it's their neural architecture that is most fascinating. Their neurons are not solely seated in the central brain, as ours are, but distributed throughout their body. A significant two-thirds of their half-billion neurons are spread along the arms; this 'mini-brain' grants them semi-autonomous capabilities, enabling the arms to search for food or manipulate objects with minimal central brain involvement. The brain still retains overall command, but this diffuse network of neurons means their arms can effectively 'think' for themselves, making decisions about how to move and interact with their environment without constantly consulting the central brain.

Octopuses also show a remarkable capacity to learn and copy the problem-solving strategies of fellow octopuses. They can negotiate complex mazes, locate hidden items, use tools and even unscrew jars. This range of behaviour hints at an inner life. Explorers encountering them in the ocean describe the almost eerie sensation of being consciously watched: a sense of being mutually aware of one another. In captivity, octopuses seem to be acutely perceptive of their confinement and the strangeness of their surroundings. Stefan Linquist, a philosopher who once studied octopus behaviour in the lab, noted, 'When you work with fish, they have no idea they are in a tank, somewhere unnatural. With octopuses, it is totally different. They know that they are inside this special place, and you are outside it. All their behaviours are affected by their awareness of captivity.' And they've been known to make some great escapes. Inky, an octopus at the National Aquarium of New Zealand, in Napier, apparently squeezed through a small gap in his tank one night, slithered across the floor and navigated a fifty-metre drainpipe that led him back to the ocean. There are reports of octopuses attempting an escape at the precise moment no one was looking, maintaining eye contact with an observer until the instant they turned away.

Octopuses also seem to have distinct temperaments; some are shy, others boisterous, some are curious and playful and others are reticent. They have been observed squirting water at bright lights above their tank, causing it to short-circuit; shooting water at people they have taken a dislike to; and playing with a floating pill bottle by shooting jets of water at it. Previously we thought this capacity to engage in activities with no obvious or immediate survival benefits was an attribute mostly of mammals.

It's humbling to learn that other life forms sense and think and experience the world very differently from us, and in ways that we cannot fathom. Our gift of sight and hearing brings the world to life, painting it in vibrant hues and enriching it with a concert of sounds. Yet our senses are inherently limited; we see only a segment of the electromagnetic spectrum and hear only a range of frequencies, and these audiovisual cues make up much of our total sensory perception. Other animals have a dazzling array of modalities beyond our own, that in some cases we are only just beginning to appreciate. Some can see in the infrared, some in the ultraviolet, some can discern the Earth's magnetic field and use it for navigating their migratory journeys, and some can perceive the tiny electrical signals produced by the muscle contraction of nearby animals.

Our view of the world is fundamentally influenced by what is permitted to us by our cognitive and sensory faculties. Against what, then, do we discern the biases in our human perspective? As it is, the only winnowing of worldviews is taking place from within the multiplicity of viewpoints we've amassed ourselves, all of which are stamped with the trademarks of human design and have been filtered through our distinct sensory apparatus. The way we perceive space and time and a world we cannot see carries the inevitable hallmarks of human perception. Would we have arrived at those same frameworks had we been equipped with a fundamentally different biology?

Without any parallels, any precedents, any peers on this planet, there exist no alternative histories of the universe, no conflicting cosmologies, only the singular story written by the one species that acquired the requisite abilities to do so. We, therefore, find ourselves in a rather insular and solitary position. Having turned our gaze to the heavens, we have asked impregnable questions about our own existence, questions that seem all the more difficult to answer alone. We have more than one reason to be intrigued by the prospects of life elsewhere in the cosmos.

magnetic fields that shielded them from assaulting radiation. But key differences set the two young planets on diverging trajectories, one becoming a flourishing oasis on which life thrived and the other devolving into a hostile, sterile desert. Forming further from the Sun and being smaller than Earth, Mars received less solar energy and cooled more rapidly; this solidified its core and shut down the magnetic dynamo that was powering its protective magnetic field. Without this shielding, its atmosphere was under continuous assault from the solar wind, and lighter atoms that constitute water were gradually stripped away, leaving a dry, barren and toxic husk. But could there exist the possibility that in its distant warm and wet past, Mars was hospitable to life?

In 2021, after journeying 300 million miles, the most sophisticated explorer we have ever sent to another world touched down in the Jezero crater of the red planet, a site that once cradled an ancient lake fed by mighty river deltas emptying their valuable cargo of sedimentary rocks into the lakebed. NASA's nuclear-powered, car-sized rover, *Perseverance*, is stacked with state-of-the-art instruments to probe the distant past for signs of microbial Martian life and survey the planet's future as a potential base for our increasing presence.

With an innovative sampling system, *Perseverance* scours and drills into the ancient lakebed of Jezero to collect rock and soil samples that it will deposit on the Martian surface for future missions to retrieve and bring back to Earth. Here we will be able to sift and study them with much more advanced technology to verify if there exist signs of past life. *Perseverance*'s high-definition zoomable camera and instruments that can detect wavelengths beyond human vision are studying ancient rocks for evidence of the carbon-based molecules that were essential for life on Earth. Such valuable data will test the compelling hypothesis that Mars's warm and wet past might have been a prime incubator for microbial life. Although there is no liquid water on Mars now, there remains the tantalising possibility that it may still lurk underground, shielded from the assault of solar radiation that bombards the planet's surface.

In anticipation of our prospective future on the planet, *Perseverance* is equipped with technology to extract the carbon dioxide that saturates the Martian atmosphere and manufacture oxygen, thus emulating the role performed by vegetation on Earth. Experiments like this can trial the potential for producing our own oxygen on Mars, which could supply breathable air for humans on future missions and may also serve as rocket fuel.

With more and more rovers trundling over the surface of Mars, it's clear we have ambitious designs on our neighbouring world, so disarmingly similar to our own and yet so strikingly different. Alongside probes sent by NASA and the European Space Agency, India's *Mangalyaan* (Sanskrit for 'Mars-craft') entered Mars's orbit in 2014, making India the first Asian nation to reach the red planet, and marking the country's first foray into interplanetary travel. China's *Tianwen-1* mission (meaning 'questions to heaven') successfully placed an orbiter in Mars's orbit in 2021 and landed a rover in the vast basin of an impact crater in the northern hemisphere. That same year, the United Arab Emirates became the first Arab nation to reach Mars with its probe *al-Amal* (Arabic for 'hope'), so named to convey the aspiration that 'Arab civilisation once played a great role in contributing to human knowledge and will play that role again'. Set for launch in the mid-2020s is Japan's *Martian Moons*

LIFE IN OUR NEIGHBOURHOOD AND BEYOND

eXploration (MMX) whose mission is to explore Mars's moons, Phobos and Deimos, and, in a world first, land on Phobos and return samples of it back to Earth.

Having charmed us since antiquity, featuring prominently in our mythology and our astrology, the faint red star we see in our night sky is no longer uninhabited; instead, a scientific armada scours its surface, sampling its iron-rich soil, hovering above its dusky terrain, monitoring its weather systems while weathering its mighty dust storms, conducting experiments and gathering intel so that one day it might be inhabited by beings of our own kind. While we might have hoped to find Martians, it seems likely now that those Martians will one day be us, and this fire star that ignited our skies and spirit may become our inaugural step en route to becoming an interplanetary civilisation.

Although Mars has been on our radar as our nearest and most Earth-like neighbour, for a long time we knew very little about the planets that lay beyond. Through the low-resolution power of telescopes we learned that Jupiter was giant and gaseous with a prominent red spot and was circled by a multiplicity of moons, and that Saturn was distinctly ornamented with an elaborate ring system and a handful of satellites. Those more distant ice giants, Uranus and Neptune, were much more mysterious, and our idea of them was hazier. To get better acquainted with these worlds beyond Mars, plans were drawn up to send spacecraft to fly past Jupiter and Saturn and, if successful, to venture beyond to Uranus and Neptune.

OUTER WORLDS

In the late summer of 1977, from the NASA base in Florida, we launched into space the *Voyager 1* and *Voyager 2* spacecraft.

Destined for Jupiter and Saturn and designed with an operational lifespan of roughly five years, they would shatter even our most audacious expectations. The *Voyagers* traversed the depths of space for decades, becoming the first entities from Earth to venture beyond the perimeters of our Sun's influence and into the mysterious unknown of interstellar space. *Voyager 1* crossed this frontier in 2012, followed by *Voyager 2* in 2018 – our very first steps into the space between the stars.

To undertake this extraordinary journey, the probes took advantage of a rare planetary alignment. A decade earlier, it had been noted that the outer planets would be arranged in a configuration that happens only once every 176 years. Here was a once-in-a-lifetime opportunity to send probes on a 'Grand Tour' of the outer solar system with unusual speed and efficiency. Leveraging this rare planetary line-up, the probes could utilise the gravity of each planet they encountered to boost their speed and adjust their trajectory and slingshot their way onto the next. Thus, the amount of fuel needed for the mission and the flight time to distant planets was cut dramatically. Both *Voyagers* visited Jupiter and Saturn, after which their paths diverged. *Voyager 1*'s encounter with Saturn's moon, Titan, sent it on a different and shorter trajectory, while *Voyager 2* would continue on to visit Uranus and Neptune – the only spacecraft ever to have done so. With the slingshot manoeuvre, the flight time to Neptune was reduced from thirty years to twelve.

Voyager 1 approached Jupiter in 1979 and, in a graphic time-lapse sequence, captured the vigorous dynamism of the planet. It revealed a turbulent Jovian atmosphere, whirling clouds of ammonia and water floating in an atmosphere of hydrogen and helium, and the iconic Great Red Spot, a spectacular storm larger than Earth that has been raging for centuries. The probe also made the surprising discovery that Jupiter has a faint ring system made of fine, dark particles, so subtle that it's only visible when the planet is backlit by the Sun. Perhaps most unexpected was the encounter with Jupiter's moon Io, which revealed towering plumes being ejected from the surface of the satellite – Io was an actively volcanic world.

More than a year and a half later, *Voyager 1* arrived at the mesmerising grandeur of Saturn, where it was to fly close to its largest moon, Titan. While the moon's thick atmosphere made it difficult to directly observe the surface, *Voyager 1* was able to gather data on its composition, density and pressure. This data led to speculation that there might exist liquid hydrocarbon lakes on Titan, marking the moon as a place we must return to. After this, *Voyager 1* was sent northward out of the plane of the ecliptic and ultimately out of the solar system.

Voyager 2 followed a different trajectory that would take it on a series of close planetary encounters, passing Jupiter after two years, Saturn in four years, Uranus in nine years and Neptune in twelve years. On its way past the Jovian system, the probe was posted on a ten-hour volcano watch that saw it witness the spectacular eruption of a volcano on Jupiter's moon Io, confirming *Voyager 1*'s suggestion that the moon harboured active volcanoes. Its encounter with Saturn saw it pass through the plane of the planet's magnificent rings, where it was intensely bombarded by tiny dust grains and had to repeatedly trigger its attitude control jets to maintain stability. Reaching Uranus yielded the discovery of ten new moons, two new rings, a tilted magnetic field and blusterous winds reaching 450 mph. Finally – and at last – encountering Neptune, it flew within several thousand miles of the cloud tops of the giant planet, discovered six new moons and four new rings, and witnessed the fury of Neptune's 700-mile-an-hour winds.

On their journey through our solar system, not only did the *Voyagers* offer a stunning window into the world of our planetary neighbours and their orbiting moons, but they also took photographs that would forever change the way we see Earth and our place in the vastness of the cosmos. Speeding out of the solar system, almost 4 billion miles from the Sun, *Voyager 1* was instructed to warm up its camera and point it back towards home for what would be the last time. It took a series of images showing Neptune, Uranus, Saturn, Jupiter, Venus and Earth, which were later assembled to form the 'family portrait' of the solar system.

The photograph that *Voyager 1* took of Earth, on Valentine's Day 1990, is one of the most iconic photographs of all time. Pointing its camera towards us, it captured our terrestrial sphere basking in a ray of light from the Sun: an impossibly tiny blue dot, about one tenth of a pixel in size. Famously coined by Carl Sagan the 'Pale Blue Dot', this photo would become a love letter to humanity, an image that Sagan, Carolyn Porco and others on the *Voyager* team felt would show us how small and acutely vulnerable our planet is in the cosmic ocean. And it did just that. From *Voyager*'s remote vantage point billions of miles away, Earth looked triflingly inconsequential – 'a mote of dust suspended in a sunbeam', adrift in the consuming vastness of the universe.

This was the ultimate 'Overview Effect' – the heightened sense of awareness astronauts report during space flight that triggers reverential awe and reflection on the meaning and place of human life. In recounting the famous Earthrise photograph he took in 1968, which shows, from a lunar orbit, a

LIFE IN OUR NEIGHBOURHOOD AND BEYOND

blue-marbled Earth rising above the horizon, *Apollo 8* astronaut Bill Anders expressed how he was '. . . almost overcome by the thought that here we came all this way to the Moon, and yet the most significant thing we're seeing is our own home planet, the Earth'. From *Voyager*'s perspective, this sensation was magnified exponentially. We had ventured billions of miles into the depths of space only to discover that the most poignant moment was looking back at our serene refuge against the staggering backdrop of the cosmos. Shortly after these photographs were taken, *Voyager 1* switched off its camera forever. It needed to conserve its energy for the long journey ahead.

That journey will take the probes into the deep space between the stars, and if they avoid any catastrophic collisions with interstellar debris, in around 40,000 years, *Voyager 2* will be 1.7 light years from Ross 248, a small, dim star nestled in the constellation of Andromeda. In the same timescale, *Voyager 1* will come within 1.6 light years of Gliese 445, a red dwarf star in the Little Dipper constellation.

The *Voyagers* carry a message for the cosmos and any intelligent life that might encounter them. On board each craft is a unique artefact: a gold-plated copper record that carries a time capsule of life on our planet. Etched in these golden records is a glimpse into our society and cultures: the breadth of languages we speak, the music we've composed, the sound of waves crashing on our shores, the wailing cries of a newborn, the hearty warmth of laughter (Sagan's), pictures of people and plants, fish and fallen leaves, seashores and sunsets and much more. There is also a message in Morse code, 'per aspera ad astra', Latin for 'through hardships to the stars', and a printed message to others in the cosmos from then US President

Jimmy Carter: 'This is a present from a small, distant world, a token of our sounds, our science, our images, our music, our thoughts and our feelings. We are attempting to survive our time so we may live into yours.'

Also broadcast on these records is our planet's location in the solar system and that of our Sun relative to fourteen pulsars, so any intelligent beings may be able to locate the origins of the probes, while a radioactive clock in the form of the uranium-238 isotope may enable them to discern how long it has been since the *Voyagers* left their home planet. In the unlikely scenario that the probes fall into extraterrestrial hands, decoding the contents of the golden record would not be the only way to learn about their architects. By dismantling the probes themselves, an inquisitive intelligence could gain insight into our civilisation's technological capabilities: our electronics, the construction materials we use and the binary logic that underlies our computations – an intimate preview into our minds and modes of thought.

As they hurtle through interstellar space, the *Voyagers* continue to transmit valuable scientific information. Equipped with a high-gain antenna, they are capable of beaming signals to Earth from up to and beyond the limits of the solar system. They are powered by radioactive plutonium, with a generator converting heat from the decay of this source into electricity to power the spacecraft. But after many decades of operation, electrical power on the probes is dwindling. In a bid to conserve energy, the probes have shut down almost all of their scientific instruments. If they remain in the range of our Deep Space Network, an array of giant radio antennas that support interplanetary space missions, we may hear from them through the 2030s. After that, they will fall silent forever. Still, with their protective shielding, the golden records on board the crafts could survive for a billion years, and the *Voyagers* will forever have the acclaim of being the first interstellar explorers, carrying with them the bold dreams and lofty aspirations of a nascent space civilisation.

OCEAN KINGDOMS

The *Voyagers'* grand tour of the solar system gave us a tantalising glimpse into the worlds of the gas giants and the many moons that circle them. The dedicated missions we've since sent have brought astonishing discoveries about the prospects of finding life in our neighbourhood. Jupiter and Saturn are composed mainly of hydrogen and helium, and if you could survive the crushing pressure as you descend deeper into their interior, eventually, you'd find that the hydrogen transitions from a molecular state (like a gas) to a metallic state. There is no definitive boundary that one could call a surface. Instead, the innermost regions of both planets are thought to consist of dense, heavy elements, many times the mass of Earth. While the planets themselves don't seem especially suitable for life, we have come to see them as entire worlds extending far out into space with their ring systems and their entourage of moons; and it's these moons that have recently piqued our interest.

The grandeur of Saturn is stunning to behold when viewed through a telescope. Possibly the most iconic and easily identifiable member of our solar system, this 'ringed planet' has a volume equivalent to 760 Earths and such astonishingly low density that the planet could even float on water. Until two decades ago, few missions had visited Saturn, although *Pioneer 11* and the *Voyager* probes briefly flew by and gave us some stunning imagery and reconnaissance to identify particular areas of interest.

It wasn't until 2004 that the dedicated space probe *Cassini* arrived at the planet. After a seven-year journey to reach Saturn, *Cassini* spent the next thirteen years there, circling the planet almost 300 times and revealing in exquisite detail its dazzling wonders, like the raging storms on the planet's northern hemisphere, housing a giant hurricane twenty times larger than anything seen on Earth. But the most dramatic and astonishing discovery from the *Cassini* mission came after data from the space probe revealed surprising details about Saturn's moons, Titan and Enceladus. These breakthroughs shattered our preconceived ideas about the places we thought were interesting for supporting life.

When *Voyager 1* sailed past Saturn's moon Titan in 1980, what it saw was an impenetrably thick, hazy atmosphere that was completely obscuring the surface. But the atmosphere had an interesting composition. It consisted of mostly nitrogen, with small traces of methane and ethane, and thus not entirely dissimilar to our own atmosphere. This hinted that interesting chemistry might be taking place below and marked Titan as a high priority for future exploration, a place we must return to.

More than two decades later, *Cassini* arrived at Titan with a precious cargo – the *Huygens* probe – specifically designed to penetrate the moon's opaque atmosphere and parachute down to the surface. After separating from *Cassini*, *Huygens* cruised solo for around three weeks to reach Titan and, once there, deployed its parachute for the two-and-a-half-hour descent. As it soared through the atmosphere, *Huygens* gathered extensive data on its temperature, chemical composition, wind speeds and pressure, and despite being buffeted by 400 kilometre-per-hour winds, touched down on the surface to become the very first probe to land on another moon in the outer solar system. Although not expected to survive landing on an unknown terrain, it continued transmitting data for several hours after touchdown.

The images sent back from *Huygens* showed a landscape that was strangely familiar – a flat, sandy plain littered with rounded pebbles and bathed in the dim orange light filtering through Titan's dense atmosphere. Most notably, the terrain showed unmistakable signs of liquid erosion, with deep gullies and drainage networks sculpting the surface – it was a world that had been shaped by flowing liquids. But with a surface temperature of −180 degrees Celsius, this liquid could most certainly not be water.

After conducting more than one hundred close fly-bys of Titan, *Cassini* confirmed that vast seas existed at its North Pole, filled not with water but with liquid methane. Although methane exists predominantly in gas form on Earth, the cold temperatures on Titan allow it to flow in its liquid state. The vast volume of liquid hydrocarbons – specifically methane and ethane – present in Titan's lakes and seas dwarf that of all Earth's oil and gas reserves combined by at least a hundred times, although they are likely to have emerged from processes very different from those that form terrestrial fossil fuels.

On Titan, methane plays the role that water plays on Earth. Methane clouds gather beneath its orange-hued skies, sending torrents of methane rain; rivers flow down from mountain tops, carving their way across the moon's icy terrain, pooling into vast lakes and seas near its poles. With stable liquid on the surface, Earth-like weather patterns and an organic-rich atmosphere, it's the most uncannily Earth-like of all the celestial bodies we've encountered.

Cassini's radar has even hinted at the possibility of a subsurface ocean of salty liquid water concealed beneath a thick icy crust that is many kilometres deep. This mix of familiar and alien features has raised the inevitable question: could life exist on Titan? In the extremely cold, methane-rich environment, life as we know it would face tremendous obstacles: for one, the chemical compounds essential for cell membranes would be frozen solid. Life would have to be based on fundamentally different chemicals, perhaps using methane as a solvent instead of water. Still, Titan's prebiotic conditions are not dissimilar to those thought to have been the catalyst for kick-starting life on Earth. As such, this moon has become a laboratory for investigating the parameters of life as we know it and life as we don't. The surprising discovery awaiting us below the thicket of its atmosphere has quashed the notion that worlds far from the Sun are unpromising destinations in the search for life-supporting environments.

The icy globe of Enceladus provided another shocking revelation. One of Saturn's moons and only a tenth the size of Titan, Enceladus has a pristine white exterior that reflects almost all the light from the Sun, making it the most reflective body in our solar system. When *Cassini* was asked to fly by Enceladus, it was evident from the very first image it captured that some interesting geology was at work in its southern polar region. Closer fly-bys showed the sheer nature and extent of this phenomenon. Enceladus was spewing massive jets of fine icy particles and water vapour hundreds of kilometres into space, some of which were thrown into orbit around Saturn forming one of its rings – the wispy E-ring – while the rest were raining back on the surface and coating Enceladus like a fresh coat of snow, preserving its reflective brilliance.

These exploding geysers suggest that beneath the thick shell of ice is an extensive liquid ocean, and that, deep within the interior of the moon, intense heat and energy are being generated and thrusting water with such force as to create this spectacular phenomenon. At such vast distances from the Sun, we were expecting frozen kingdoms, yet this extraordinary moon is producing enough heat to keep water in its liquid state and prevent it from freezing. How was it doing this? The extensive data *Cassini* collected has shed more light.

As a miniature moon orbiting a massive planet, Enceladus experiences the immense pull of Saturn's gravity, and, as with the tides in our oceans, this gravitational tug causes the moon to stretch and deform as it orbits. The tidal friction then generates heat within the moon. This tidal heating is further amplified by a gravitational resonance with another of Saturn's moons, Dione. We've also learned that Enceladus probably has a highly porous core, and this allows cold water from the subsurface ocean to seep deep into the moon's interior, where it interacts with the rocks and absorbs heat, helping to sustain Enceladus's thermal activity.

This internal heat, in turn, creates hydrothermal vents on the ocean floor, releasing hot, mineral-rich water. Similar to those found on Earth, these vents spew heated water that rises and eventually erupts through enormous tiger-stripe fissures in the icy crust, creating the spectacular plumes that *Cassini* observed. *Cassini* even flew through the plumes, sampling their composition and finding organic molecules and salts, thus strengthening the case that Enceladus's ocean contains some key ingredients for life as we know it. This startling discovery has positioned what had seemed to be a small, unassuming ball of snow as a leading contender in the search for life beyond Earth.

Where we were previously limiting ourselves to the Goldilocks zone around middle-aged stars as the most likely place for habitability, these findings have compelled us to significantly widen that window. Not only can potential habitats for life exist very far from the warmth of a star, hidden in deep underground oceans heated by tidal forces, but planets orbiting stars are not the only places; their moons are also viable candidates.

Having spent thirteen years giving us exclusive and in-depth coverage of the world of Saturn and its moons, *Cassini* eventually ran out of the fuel that was primarily used to adjust its course. To avoid any undesirable consequences, a decision was made to execute a controlled end to *Cassini*'s mission. As one of the longest and most successful endeavours in the history of space exploration, *Cassini*'s grand finale was also a spectacular affair.

Diving nearly two dozen times through the narrow gap between Saturn and its innermost rings, the probe beamed back incredible images and valuable data on the material composition of the rings to help us understand their origins. After the last dive came the final plunge; *Cassini* was programmed to head straight for Saturn and intentionally plough into the gas giant.

On 15 September 2017, *Cassini* made its final, fateful descent into Saturn's atmosphere. As it plummeted into the planet, *Cassini* continued to gather extensive data, sampling the atmosphere's composition and measuring the planet's gravitational and magnetic fields with unprecedented precision. The probe transmitted this information back to Earth until the increasing atmospheric pressure overwhelmed its ability to keep its antenna pointing to its home planet. Eventually, contact was lost and the probe fell silent. *Cassini* had been vaporised in Saturn's atmosphere, a dramatic but deliberate manoeuvre to avoid the unlikely scenario that it may instead collide with one of Saturn's moons. That these icy worlds could potentially support life made this a necessary precaution, so that contamination from Earth-borne microbes that might have lingered on the probe would not inadvertently compromise future investigations. For a probe that had spent its operational lifespan dedicated to the Saturn system, it was a poignant and fitting end.

Cassini's groundbreaking findings at Enceladus have sparked a fascination with these enigmatic 'ocean worlds'. Europa, one of Jupiter's four largest moons, is strikingly similar to Enceladus. Beneath Europa's icy exterior lies a global ocean of salty water, potentially holding more liquid than all of Earth's oceans combined. This vast ocean is shielded by an incredibly dense sheet of ice, frozen harder than granite and possibly tens of kilometres thick. Like Enceladus, Europa experiences powerful tidal forces from its host planet, Jupiter, and this gravitational tug-of-war generates internal heat, keeping the ocean liquid and potentially powering hydrothermal vents on the ocean floor. These vents could provide the energy and chemical building blocks necessary to support life, making Europa another prime candidate.

Our probes have circled and landed on many worlds within our reach and shown that, while there are no apparently advanced life forms, no irrigation network of canals and no signs of an alien civilisation, there is often more than meets the eye. The worlds we previously thought to be desolate and dry, so far removed from the life-giving power of the sun as to be inertly frozen and hostile spheres, may not be quite so barren. We have found that conditions conducive to the emergence of life may be lurking deep within their interiors, and thus microbial life may be hidden from our view, thriving in a 'dark biosphere'.

Future missions will reveal whether life has indeed found a way on worlds that, superficially, appear so unforgiving. As we extend our reach into the solar system, we are getting ever closer to answering the question: does life exist elsewhere in our cosmic neighbourhood, or has it been a phenomenon exclusive to this third planet from the Sun?

WORLDS BEYOND OUR SUN

For worlds close to home we can send our robotic scouts to survey the landscape. But we are also aware of the insignificant extent of our solar system – and we want to know if there is life beyond our Sun.

For centuries, people have speculated that there might be planets outside our solar system, planets orbiting other stars. But we only confirmed this experimentally in 1992. At the Arecibo Observatory in Puerto Rico, radio astronomers Aleksander Wolszczan and Dale Frail were studying a 'dead star' or pulsar – the very dense remnants of a massive star that exploded in a supernova. Such stars emit regular pulses of radiation, like a lighthouse beacon, and, after observing these pulses, the astronomers noticed irregularities in their timing. This suggested that the star was being gravitationally tugged, causing it to wobble and pulsate in these slightly irregular beats. After careful analysis, they concluded that the root cause of this was planets in orbit around the star, at least two of them. This discovery confirmed the first detection of an 'exoplanet' – a planet residing outside our solar system.

Since then, as our telescopes have become more powerful and our techniques more refined, the list of confirmed exoplanets has very rapidly grown longer and longer. Operational from 2009 to 2018, the Kepler Space Telescope was a pioneer in exoplanet hunting; its chief mission was to stare at a fixed patch of sky in the Cygnus constellation and continuously monitor the brightness of over 150,000 stars. It discerned the existence of an exoplanet using the 'transit method' – looking for very subtle dips in a star's brightness caused by a planet momentarily passing in front of it (transiting). And it was extraordinarily successful, discovering thousands of confirmed exoplanets and thousands more potential candidates.

With Kepler, it became abundantly clear not just that planets were in plentiful supply in our galaxy, but also that there was incredible diversity: they come in various sizes and masses, can be rocky or gaseous and everything in between, have variable or no atmospheres, and orbit their parent star in a multitude of configurations. If we might have harboured any lingering illusions that our solar family was unique, this data emphatically quashed them. We now had a growing catalogue of stars attended by a retinue of planets. The sheer multiplicity of worlds out there also fundamentally shifted the probability of finding life, even sparking searches for Earth 2.0.

Kepler's successor, TESS (Transiting Exoplanet Survey Satellite), took over in 2018 and was designed to survey almost the entire sky, but with a special emphasis on bright, nearby stars, and Earth-sized or super-Earth-sized planets that might be circling them. TESS intends to draw up a list of exoplanets and hand them over for further scrutiny to the most powerful space telescope we have ever built, the James Webb Space Telescope (JWST).

Launched in 2021, the James Webb telescope was sent a million miles from Earth to a stable point, where it can maintain its very cold operating temperature and adequately shield itself from the Sun, Earth and Moon. The telescope has highly sensitive instrumentation and can observe deeper in the infrared, allowing it to see much fainter objects than its predecessor, the Hubble Space Telescope, and also see much further back into the history of the universe. It is especially well equipped to analyse an exoplanet's atmosphere, searching for the telltale signs of life or 'biosignatures'. Life as we know it invariably alters the atmosphere of its host planet. If a scientist on a distant world were pointing a sufficiently powerful telescope in the vicinity of our solar system, they might identify a planet, in the temperate zone of a star, with a characteristically distinct atmosphere (containing oxygen, methane and water vapour), signalling that something interesting was at play.

The Webb telescope has the functionality to do the same with exoplanets and is expected to revolutionise the field. When an exoplanet passes in front of its star, a fraction of starlight filters through the planet's atmosphere before reaching the telescope. Different molecules in the atmosphere absorb specific wavelengths of light, leaving unique 'fingerprints' in the observed starlight. By analysing this filtered light, we can discern which molecules are present in the planet's atmosphere. The presence of these molecules in specific combinations, especially those that are difficult to explain through non-biological processes, could be compelling evidence for the possible existence of extraterrestrial life. Given this capability, it's clear that

LIFE IN OUR NEIGHBOURHOOD AND BEYOND

our search for signs of life beyond Earth is heading towards a step change.

With data from ground- and space-based telescopes, the current count of confirmed exoplanets is over 5,000 and is constantly growing. Working outwards from these numbers, it's estimated that billions of exoplanets exist in our galaxy alone. Among this planet inventory are: gaseous planets the size of Jupiter orbiting very close to their host star (hot-Jupiters); rogue planets wandering lonesome in space, not orbiting any star at all; rocky planets much larger than Earth (super-Earths); planets orbiting dead stars; planets in highly eccentric orbits around their host stars; planets tightly bound to their stars and those loosely circling them; planets where it rains diamonds or molten iron or liquid rubies or sapphires or molten glass; and even planets circling multiple stars.

Nature has crafted such distinct and diverse worlds out there that the wildly fictional ones we once dreamed up are turning out to actually exist. For instance, Kepler-16b was the first planet found to have two suns in its sky – eliciting comparisons to Tatooine, the fictional desert planet that was home to Luke Skywalker in *Star Wars*. It orbits a binary star system, and, while it is unlikely to be suitable for life, another planet in that system, Kepler-453b, resides in the habitable zone. If life exists on that planet, its residents can witness the splendour of twin sunsets. Since the discovery of Kepler-16b, we've observed even more complex arrangements, including those where planets orbit three-star and four-star systems.

Another interesting phenomenon is planets that are tidally locked to their stars. These are worlds where one side forever faces the star and is bathed in perpetual daylight, while the opposing side is engulfed in eternal darkness; on one hemisphere, the Sun never sets, and on the other, the Sun never rises. These extremes generate incredibly hot temperatures on one half and a penetrating cold on the other. If such planets have a suitable atmosphere or liquid oceans or extreme supersonic winds, they may be able to distribute heat effectively to regulate their temperatures. And while they present a significant challenge for life, it is theorised that there may exist a region between the day and night sides – the twilight zone – where more moderate temperatures, and therefore life, might be possible.

Of all the exoplanets discovered thus far, the oldest is 12.7 billion years old, meaning it formed only a billion years after the Big Bang. Such ancient worlds hold great knowledge. If we could study their composition, we might gain better insight into the early stages of the universe and just when the raw materials considered essential for life became widespread.

What is the closest exoplanet to us? Remarkably, the nearest planet orbits our closest star, Proxima Centauri, located 4.25 light years away – in cosmic terms, this is just down the road. Discovered in 2016, Proxima Centauri b is just a little over the mass of the Earth and orbits in the habitable zone, which means it could potentially sustain liquid water on the surface. But the planet might be tidally locked to its star, making conditions very extreme. Also, the star itself is small and low mass but also quite volatile; it spontaneously discharges violent solar flares, unleashing torrents of radiation that could create an environment very hostile to the emergence of life. Yet, despite this, we have a candidate planet circling our nearest star that may, in principle, be habitable. And we have ambitious plans for it.

Breakthrough Starshot is a bold initiative to send a fleet of miniature, ultralight space probes to the Proxima Centauri star

LIFE IN OUR NEIGHBOURHOOD AND BEYOND

system within the span of a generation. If we can attain speeds up to 20 per cent of the speed of light, these probes could reach this star system in around twenty years. The most plausible idea to emerge is a fleet of nanocrafts attached to light sails, that are propelled to a fraction of light's speed by a powerful array of ground-based lasers. This futuristic concept builds on early successes like Japan's *IKAROS* mission, which in 2010 became the first spacecraft to demonstrate solar sail propulsion in interplanetary space. *IKAROS* unfurled a fourteen-metre-wide sail and used the pressure of sunlight to accelerate itself and power its onboard systems on its journey towards Venus.

Breakthrough Starshot aims to take this principle much further. Each tiny probe will be fitted with a host of miniaturised equipment, the goal being a postage-stamp-size, gram-scale computer chip complete with power supply, navigation, cameras, thrusters and transmitters. At the same time, advances in nanotechnology appear promising for fabricating materials that could be used as the sails, which must be extremely lightweight, of the order of grams, and very thin but metres-scale in length. A swarm of a thousand of these miniature space probes are likely needed to compensate for collisions with interstellar dust. Unsurprisingly, executing such a daring enterprise will push the limits of our technologies, but, despite daunting challenges, it is thought to be achievable using current and future projections.

The ultimate goal of Starshot is to bring interstellar travel at a fraction of light speed to within humanity's reach. If viable, this would be a historic milestone with significant consequences. For one, it will help us escape the 'incentive trap' – the dilemma that a ship launched now will be overtaken by

A BRIEF HISTORY OF THE UNIVERSE

a faster one launched years from now (due to technological progress), so why bother? Our farthest spacecraft, *Voyager 1*, is currently 15 billion miles away. A sail travelling at one fifth the speed of light could overtake it in less than five days – a journey that took *Voyager 1* nearly half a century. Yet, once we reach this threshold of around a fifth of light speed, there will still be an incentive to launch, because future missions will not beat us to the destination.

In the coming centuries, with these technological developments afoot, our search for life will extend to the wider neighbourhood, to distant suns and their planets scattered across our galaxy. We will then have a much clearer idea of just how populated the cosmos is.

CHAPTER 11

Decoding Messages from the Cosmos

In our galaxy alone, we think there exist at least as many planets as there are stars. With up to 400 billion stars in the Milky Way, this translates to a staggering population. Sky surveys with the Kepler telescope show that around 300 million of these planets might potentially be habitable – a suitable distance from their star to sustain liquid water on the surface. And these projections are for our galaxy alone, which is only one of among an estimated 2 trillion galaxies. Our nearest neighbour galaxy, Andromeda, is even more populated than ours. The number of planets is only likely to proliferate.

Considering this abundance, it seems statistically improbable that life hasn't materialised anywhere else in the universe. It appears almost absurd that we are alone. Yet, we have not heard from anyone; no radio transmissions have been intercepted of alien communications; no intentional broadcasts from another star have been detected; no evident artefacts of interstellar explorations have been observed. The silence of the stars that once stirred our poetic souls has become disconcerting. Where is everybody? This is the sentiment that physicist Enrico Fermi expressed over a lunchtime conversation with colleagues in the summer of 1950, and it's become known as the Fermi paradox.

THE GREAT SILENCE

The question appears deceptively simple, but its underpinning arguments are strongly compelling. Our galaxy is 13.6 billion years old, and if there exist advanced civilisations out there, we expect that some will be behind us while others will be far ahead of us. We became a technological civilisation not so long ago, and yet, in that short span of time, technology has grown so explosively fast that the tools we wielded only decades ago now appear primitively archaic. From this exponential growth, we might extrapolate that a civilisation far ahead of us is unimaginably advanced in comparison. Even with modest rocket technology, an imperially motivated civilisation could, in principle, spread throughout the Milky Way within a few tens of millions of years.

We can reason that they may dispatch spaceships with the intent of 'terraforming' other planets to make them suitable for habitability, beginning with those in their immediate vicinity before expanding to more distant star systems. But the more plausible scenario is that the civilisation sends out an uncrewed self-replicating probe, called a von Neumann probe, after the mathematician John von Neumann, who theorised it. This probe, guided by a sophisticated artificial intelligence, would be capable of travelling at a fraction of the speed of light. Programmed to land on a suitable host site – a planet or moon or asteroid – it would mine the raw materials of that site to create replicas of itself, which would then be launched in various directions to allow for exponential expansion across the galaxy.

The Milky Way stretches approximately 100,000 light years across. With a self-replicating spacecraft that can travel at 10 per cent of light's speed, an advanced intelligence could conceivably colonise every planet within our galaxy over the span of a few million years. Even if that speed were limited to a more modest 1 per cent of the speed of light, colonisation of our galaxy might take around 20 million years instead. The exact numbers, of course, are not very relevant because astronomical time scales are so vast. Self-replicating probes offer the advantage of rapid colonisation, which makes the Fermi paradox even more puzzling, but the paradox persists even if such probes are not factored in.

Even if interstellar expansion is exceptionally slow, a civilisation that arose millions of years ago would have had ample opportunity to spread through the Milky Way, in what is, on a cosmic scale, a relatively short time. And if our galaxy has been around for so long, admitting such a large window for a species to emerge, evolve and extend its reach, why haven't we encountered any signs of them? Where are the remnants of their empires, the artefacts of their engineering, the footprints of their exploratory missions? It would have taken just one alien civilisation to pierce through the 'Great Silence'. Yet, we are met with a haunting quietude.

Where, indeed, is everybody? One popular hypothesis that has received increasing attention is the notion of the 'Great Filter'. This idea suggests that there exists a critical bottleneck – an obstacle (or set of them) that must be surmounted for a civilisation to expand across the cosmos and become discernible.

We haven't heard from anyone or seen any visible signs of their presence because most, if not all, were unable to negotiate this hurdle. From life's singular trajectory on Earth, we have come to appreciate the series of improbable odds overcome to get to where we are: the transition from non-life to life, from simple life to complex life, the evolution of intelligence that can wield tools, and the emergence of a technological civilisation. We have yet to discover the challenges that lie ahead as we're still apprenticing as a space civilisation. But the Great Filter speculates that at least one of the steps en route to galactic colonisation may be insurmountable, filtering out almost all emerging civilisations and precluding them from making contact or being visible to others. Even the discovery of the most primitive organisms on Mars would be a spectacular breakthrough; it would indicate that life's emergence from inanimate matter is not so rare, and may shift the location of the Great Filter further along the chain.

In this context, then, two possibilities arise for us. One is that we have already overcome this imposing obstacle, and the Great Filter is behind us. On the other hand, there is the disquieting prospect that this filter is still ahead of us. Having surpassed a critical threshold where we now have the means to annihilate ourselves, we must ask: Do all such civilisations carry an intrinsic self-destruct mechanism? Have we not heard from them because they inadvertently pulled the trigger of their own demise? Or perhaps extinction-sized catastrophes are responsible for this silence: asteroid bombardment, supervolcanism, a nearby exploding star, or other exotic cosmic phenomena that might sever the evolutionary chain leading up to intelligent observers. While large swathes of life on Earth were also wiped out in similar catastrophes, perhaps our planet has been unusually lucky, and these cosmic resets are much more frequent and more severe than our planet's history has led us to believe.

There are other possibilities. Maybe extraterrestrials are just not interested in colonising, or maybe they are lurking stealthily and observing us like animals in a zoo, waiting for the opportune moment to make contact (Zoo hypothesis). Maybe their fierce instinct for self-preservation has instilled in them a deep suspicion of others, and they are resolute on annihilating any life with which they come into contact, to preserve themselves (Dark Forest theory). Or maybe they've transformed into something else entirely with powers and purposes that are far beyond our comprehension (Transcension hypothesis). However, these explanations are not thought to be sufficient on their own to solve the paradox because they make assumptions about motivations and behaviour that might well apply to some extraterrestrial civilisations, but are unlikely to extend to them all. It would only take one rogue civilisation to break the mould and make itself known.

An alternative explanation for the Great Silence proposed before the discovery of exoplanets is the Rare Earth hypothesis, which posits that our planetary status may not be as mediocre as we think, and hence, intelligent life may well be exceptionally rare. It enumerates the great many factors that had to align to produce complex life on Earth; a stable, long-lived star that's neither too active nor too hot, residing in a habitable zone of the galaxy; a planet orbiting at the right distance to sustain liquid water; a large moon to stabilise our planet's wobbles and moderate its climate; global-scale climate events that coincided with the flourishing of diverse life forms like in the Cambrian explosion, and so on. But the hypothesis is thought to be very 'Earth-centric' in its approach, placing too great an emphasis on life requiring the exact conditions and sequence of events as played out on Earth.

SCALES OF CIVILISATIONS

If intelligent civilisations are somewhere out there in the swirling cities of stars, what might they be like, and how could we discern their existence? To imagine civilisations that are thousands or millions of years ahead of us, we have to take the flights of fancy that one might associate with science fiction. Having taken our first tentative steps into the cosmos a mere century ago, we are still a very young space civilisation with a future that remains highly uncertain. We cannot even confidently assert what this future will look like a few hundred years from now, let alone thousands or millions of years, because the exponential progress of technology has rendered such projections impossibly challenging. As the science fiction writer Arthur C. Clarke famously said: 'Any sufficiently advanced technology is indistinguishable from magic.' Indeed, if a superior intelligence has learned to operate technology based on principles that we have yet to discover, these could be so far beyond our comprehension that we could be staring at them without recognising them.

Hence, when thinking about how we might discern this superintelligence, we have to be rather audacious. One imagined scenario came from Soviet astronomer Nikolai Kardashev in 1964. Kardashev looked at the power that an advanced civilisation would need to transmit information across large distances in space, as well as our hopes of intercepting that communication. He came up with a framework for categorising a civilisation based on how much energy it can harness. On the very first rung on this ladder is a Type 1 civilisation – a planetary civilisation that has learned to tap into the energy available on its home planet, which would include the energy from its star and the energy from the internal heat of the planet. We're currently not yet a Type 1 civilisation because we have not fully harnessed the energy of our planet, and so are estimated to be around 0.73 on the Kardashev scale.

The next level is a Type II or stellar civilisation that has learned to exploit the total energy output of its star to keep pace with its escalating energy demands. The energy typically reaching a planet from its star is a very small fraction, and, once that resource has been exhausted, the next stage would be to go straight to the source and directly extract the energy of the star itself. Such a civilisation might employ the hypothetical megastructure proposed by physicist Freeman Dyson – called the Dyson sphere – a megastructure that partially or fully encloses a star to capture its energy, maybe using an orbiting swarm of satellites that intercept the energy and collect it. A star encircled in such a way will appear to have lower luminosity, so we may observe a dip in the light we receive from it, which a transiting planet or other natural phenomena cannot explain. By searching for anomalous signals from stars, we may be able to discern unusual behaviours that we cannot attribute to other known phenomena, which may hint that something interesting is going on.

Further ascending this ladder of energy usage, we meet a Type III civilisation – a galactic civilisation that can control energy on the scale of an entire galaxy. This would require technologies far beyond our current comprehension, allowing the civilisation to tap into the energy of multiple stars and perhaps even harness the power of black holes. They may even have the ability to counteract expansion and retain a conglomeration of galaxies within closer confines to make it easier for them to exploit their energy.

Given the extent of their dominion over extensive scales, Type II and Type III civilisations would be transmitting large volumes of information, and doing so continuously. Kardashev concluded that we were more likely to deduce the presence of large-scale civilisations broadcasting at this level.

In all of this we've been thinking about a much larger scale and speculating on what the future millions of years of technological advancement might usher in for a civilisation in search of energy or other resources.

But projections for mastery exist at the other end of the scale, too. John Barrow proposed that we look at the possibility of alien civilisations and their ability to manipulate matter, but at the smallest of scales. Rather than an outward expansion, he proposed that they may try to master the underlying workings of the universe and thus control and manipulate it at its most fundamental. This inversion of the Kardashev scale counts down from a civilisation of Type I-minus, which has the ability to assert control over and manipulate objects at the scale at which we exist. Then you progressively descend further and further down: Type II-minus can tinker with genes, Type III-minus can manipulate molecules, Type IV-minus can engineer/manipulate atoms, Type V-minus controls atomic nuclei, Type VI-minus can manipulate elementary particles and Type Omega-minus has the capability to modify spacetime itself.

In this model and for these lower levels, it's not obvious that we would be able to distinguish an alien life form from the workings of nature. For instance, if a civilisation has acquired Type Omega-minus status and thus learned to manipulate spacetime itself, then they are pretty much integrated into the fabric of our reality.

Many of these speculative notions may well seem extraordinary and outrageous. But they serve, at the very least, to broaden the province of our imagination. The task is to conceive how intelligent life might have played out elsewhere in the cosmos and how a civilisation that may be millions of years ahead of us might exhibit itself. And that is manifestly challenging.

SEARCHING FOR AND MESSAGING EXTRATERRESTRIALS

These hypothetical frameworks help guide the search for extraterrestrial intelligence by suggesting what kinds of signals or technosignatures we might look for. How could we possibly begin? An earnest Search for Extra-Terrestrial Intelligence (SETI) set out in the 1960s, when Frank Drake, often considered the 'father of SETI', carried out the first modern SETI experiment, Project Ozma. He used a radio telescope to search for signals from two nearby stars and is also famous for developing the Drake equation, a formula for estimating the number of detectable civilisations in our galaxy, which continues to frame SETI research. In 1959, physicists Giuseppe Cocconi and Philip Morrison published a landmark paper on how to search for interstellar communications, which also proposed using radio waves to seek extraterrestrial signals. This paper laid the scientific foundation for modern SETI. Since then, SETI has evolved to encompass various approaches, including radio and optical searches for signals and technosignatures (such as Dyson spheres and other large-scale astroengineering projects) that an alien intelligence might have employed to extract the energy of its star.

An ideal medium for transmitting information across the vast space between stars is radio waves; they journey at light speed and are unimpeded by interstellar gas and dust. Hence, for over half a century, our primary method has been to use giant antennas to eavesdrop on any radio signals others might be transmitting. Our focus, in particular, is on signals containing telltale features, hinting that their origin may be a purpose-built transmitter, and not a natural source of cosmic noise. The radio emissions we are searching for are 'narrow-band signals' – they span only a small segment of the radio spectrum, typically just a few Hertz wide, but condense a lot of energy into that short radio space and are thus more likely to grab our attention. Such signals would be like tuning the radio, and, for the most part, hearing a lot of static and then, all of a sudden, hearing a squeal. As they are generally the easiest type of signals to find, a transmitting alien vying for the attention of others in the cosmos is likely to use them as part of their signalling. Natural cosmic sources like pulsating stars or interstellar gas are also emitters of radio waves, but they are usually spread out over the radio spectrum.

If ardent radio astronomers are indeed out there practising their craft, there are also some natural sources of narrow-frequency signals that we may expect them to be aware of, like a universal radio station that they would know to tune into. In the microwave part of the spectrum lies a particularly compelling signal from the natural emissions of hydrogen, resonating at 1420 MHz, which corresponds to a wavelength of 21cm and is famous in astronomy circles as the 21cm line. (This is due to the quantum nature of electrons in the hydrogen atom, where an electron moves to a lower energy state and emits a radio photon with a frequency of 1420 MHz.) Since hydrogen is the most abundant atom in the universe, this is a scientifically important radio signal and one that we would expect an advanced race to be aware of.

If you combine a single atom of hydrogen with a single atom of oxygen, you get a molecule called hydroxyl, which is also present in the interstellar medium and produces emissions at a frequency of 1666 MHz, corresponding to a wavelength of 18cm. These two emissions – hydrogen and hydroxyl – are both interesting for understanding the composition and structure of the vast clouds of interstellar gas, and may also be of interest to any enterprising astronomers out there. On our planet, both of these emission regions (specifically 1400–1427 MHz and 1660.6–1670.0 MHz regions) are part of a 'protected spectrum', meaning they are of such great relevance for astronomical purposes that terrestrial transmissions in these frequency ranges are forbidden. We may speculate that this rationale has also been adopted by extraterrestrial radio astronomers and that they may intentionally choose not to broadcast in this region to avoid polluting what is a valuable channel for scientific information.

But the frequency range bracketed between these two scientifically prominent signals is remarkably quiet, and hence an enticing prospect for communicating. This lull in the spectrum means that civilisations broadcasting in this frequency range would be somewhat easily discernible. The former head of the SETI programme, Bernard Oliver, called this band of frequencies (between water-relevant lines) where we might converge and communicate with others in the cosmos the 'waterhole', a place 'where species have always gathered'. Many of our radio searches in SETI have thus been devoted to this waterhole and also tuned to other 'magic frequencies', such as values that are twice the hydrogen line or the mathematical constant of pi multiplied by the hydrogen line.

In 1977, the Big Ear radio telescope in Ohio, a distinctly designed radio telescope that operated for over two decades as the longest-running experiment to search for alien intelligence, picked up a rather intriguing signal. Jerry Ehman, a volunteer astronomer working on the project, was stunned to see a strong, narrowband radio signal at a frequency tantalisingly close to the hydrogen line that lasted for the full seventy-two seconds that the telescope was pointing in its direction and had an intensity that far surpassed the background noise. Astounded by the signal's characteristics, Ehman circled it on the computer printout and enthusiastically scribbled 'Wow!' in red pen, thus forever christening it the 'Wow! Signal'. It appeared to come from the direction of the constellation Sagittarius. While it ticked a number of boxes for what we might expect from an intentional extraterrestrial signal, to be regarded as credible it was critical that the signal be replicated. Since then, many attempts have been made to detect the signal again, but to no avail. A multitude of explanations have been put forward to account for it, including a long list of cosmic phenomena – with a recent one claiming that emissions from a star may have energised a cold gas cloud, causing a sudden surge in brightness. But, as yet, there is no conclusive confirmation of the source of this signal, and, since it was never heard again, it remains a mysterious anomaly.

DECODING MESSAGES FROM THE COSMOS

Our ongoing search for anomalous radio transmissions continues with the Allen Telescope Array: forty-two smaller telescopes arranged over a 1km region in the Cascade Mountains, 300 miles north of San Francisco. Working together to survey large portions of the sky, this array listens out for a large range of frequencies, thus casting a wider net for any potentially alien signals. The telescope has paid close attention to nearby stars that are thought to be habitable and also pointed towards the galactic centre – home to a high density of stars and the possibility that older, more advanced civilisations might reside there. Operative almost round the clock, with the capacity to tune into many more frequency channels than before and to keep watch over different areas of the sky all at once, it has been a game-changer for SETI searches. Recently, we've been using this telescope array to home in on potential radio bursts from some of the most promising exoplanets discovered close to home, for instance, the TRAPPIST-1 system –a small star about forty-one light years from Earth with a phenomenal seven rocky planets circling it, some of which are in the proverbial habitable zone.

Our most thorough and extensive search for intelligence beyond our planet is through Breakthrough Listen, from the same foundation that plans to send a probe to the nearest star system. Surveying the closest million stars to Earth, the project systematically scans the entire plane of the Milky Way, where the majority of the galaxy's stars congregate, and even listens for signals from beyond our Milky Way to a hundred nearby galaxies. To this end, it employs the most powerful telescopes that offer complementary information and combines this with cutting-edge, sophisticated algorithms to sift through the vast sea of data and identify potentially interesting signals that might be of artificial origin.

Beyond tuning into the cosmic airwaves and sweeping through the innumerable frequencies at which others might be transmitting, how else might we discern their presence? Until recently, our search for signals from extraterrestrials were almost exclusively focused on radio waves. But now an innovative alternative is set to broaden the search, based on the long-held speculation that advanced alien societies might tap into powerful lasers for communication. The initiative LaserSETI is searching for brief pulses of laser light from other star systems. Intensely powerful and acutely focused, lasers can travel vast distances with minimal loss of signal, making them an alluring choice for interstellar communication. Light's ability to transmit significantly more data per second than radio – around 500,000 times more – gives it a commanding advantage. Lasers can be modulated to carry large amounts of information, with the potential to encode simple messages or even complex data. The high bandwidth this entails may be especially valuable for a spread-out extraterrestrial race wishing to communicate with scattered colonies.

Unlike traditional telescopes that focus on a narrow patch of sky, LaserSETI uses wide-field optics to observe a large portion of the sky simultaneously and is extremely sensitive, capable of detecting even faint laser pulses from distant stars. The eventual goal of LaserSETI is to have a network of observatories around the world such that we can monitor a vast portion of the sky all of the time and complement conventional radio search efforts by exploring a different part of the electromagnetic spectrum.

However, these approaches are still human-centric. We tend to assume that extraterrestrial civilisations will use technology similar to ours, such as radio waves or lasers, for communication.

DECODING MESSAGES FROM THE COSMOS

This may well not be the case, especially for civilisations far more advanced than ours. In an effort to be less anthropocentric, other approaches are emerging where the intention is to be on the lookout for anything that is unusual or unexpected. The challenge, of course, is this: how does a newcomer on the galactic scene, a civilisation in an embryonic phase of its space exploration, envisage what a dramatically more advanced race would do and how they might make themselves known? It is here that the earlier frameworks, such as the Kardashev scale, become valuable for stepping beyond our limited frontiers. This is the approach of Dysonian SETI, which emphasises pure observation and the analysis of anomalies without making assumptions about the technology behind them. It's concerned with looking for large-scale structures, like a Dyson sphere, or other unexplained anomalies that we cannot attribute to natural sources and that may hint at an artificial origin.

An alternative is that instead of listening for messages from 'them', we send messages into the cosmic void in the hope that a far more advanced civilisation may intercept them, despite the limitations of our signalling. This is the province of METI (Messaging Extra Terrestrial Intelligence), and, as one might expect, it has its fair share of supporters and detractors. The term METI was coined by the Russian scientist Alexander Zaitsev, who thought that if no one had yet penetrated the Great Silence, then perhaps we should do it, in effect 'bringing to our extraterrestrial neighbours the long-expected annunciation "You are not alone!"'

METI is tasked with creating and transmitting interstellar messages, with the initial target being nearby stars. We've been doing this since the first Arecibo message in 1974. It contained basic information on humanity (the atomic numbers of the elements that make up our DNA, our size, what we look like and how many of us there are), a graphic of our solar system and the home planet. But it was meant primarily for the purpose of showing that we have the technology to send messages rather than a serious attempt at communication. We are not expecting an alien civilisation to be able to decode the message or its meaning. The message was sent to a globular cluster of several hundred thousand stars in the constellation of Hercules – a distance of 25,000 light years away – so it is still only a very small fraction of the way on its journey. Since then, we've targeted star systems that are much closer. In 2017, we sent a message containing thirty-three pieces of music to a potentially habitable exoplanet – a super-Earth almost three times the mass of our planet and orbiting the red dwarf star Luyten's Star, located twelve light years from Earth. Along with transmitting across the cosmic airwaves, we've also sent physical messages on board *Voyager 1* and *Voyager 2* which carry the golden records, and their predecessors, *Pioneer 1* and *2*, which carried similar information.

Some think it unwise to reveal potentially sensitive information about our whereabouts that could endanger our existence if it fell into the hands of aliens with malicious intent. Among the prominent critics was the late Stephen Hawking, who thought it foolhardy to transmit our whereabouts and considered it more prudent to 'lay low'. He outlined the deleterious consequences that pepper our own history when contact has occurred between civilisations with markedly different levels of technology. But as some have pointed out, since the advent of radio we have been unintentionally broadcasting our existence for over a century to any intelligent being out there with powerful enough equipment to detect it. Others have alluded to specific hypotheses like the Zoo hypothesis and argued that if, indeed, we are being watched like animals in a zoo, then perhaps we should extend an invitation to our secret observers and elicit a response as a way of testing the hypothesis.

What if we were to one day observe a signal that has all the characteristics of an artificial origin? The first task would be to unequivocally confirm that it has come from an extraterrestrial source and rule out any possibility of natural cosmic or terrestrial origins. Our international protocol states that no reply to the signal be made without consultation with a representative international community, such as the United Nations. It would profoundly alter our cosmic perspective. Societally, it might evoke a spectrum of responses, from excitement and curiosity to alarm and existential anxiety. Scientifically, it would unleash a whirlwind of enquiry. Maybe this alien life is not carbon-based, doesn't use RNA/DNA as its blueprint, and perceives the world through an entirely novel, even unfathomable sensory mechanism. Can we converge on a common language to exchange scientific information with them? How different is their story of the universe compared to ours? What do they make of the fundamental components of the universe, and are they governed by quantum mechanics? Is the book of nature really written in the language of mathematics, or is that the language we invented to help us parse the peculiarities of nature?

Alternatively, given the distances involved and the possible barriers to life and consciousness and interstellar travel, even if there is life elsewhere we may never have the opportunity to meet. For all practical intents and purposes, then, there exists

A BRIEF HISTORY OF THE UNIVERSE

the possibility that we might always be alone, not because life hasn't emerged on other worlds but because our chances of encountering them are vanishingly small. Perhaps too many stars have to align for such a rendezvous.

What would that mean for us? Arthur C. Clarke said: 'Two possibilities exist: either we are alone in the Universe or we are not. Both are equally terrifying.' For an inherently social species with a deep desire to connect, the prospect of being alone in this vastness may be the more unsettling. It would also greatly raise the stakes. If the experiment called life has not been repeated (as far as we can see), then what transpired on Earth takes on eminent significance. Even if we are not alone, the story unfolding on this pale blue dot is almost certainly without parallel.

Our Future in the Cosmos

CHAPTER 12

A New Era

Ever since our ancestors traced the paths of the twinkling lights across their night skies, our collective gaze has been drawn to the heavens. We've marvelled at the wonders of the universe and at the capacity of our intellect to deduce the laws that appear to govern its workings. Our telescopes have peered into the depths of space, revealing its staggering scale and the colossal structures that populate it; our rovers have scoured the terrains of other worlds, and particle colliders have given us a glimpse into the perplexities of the quantum world, where subatomic particles and the forces that bind them sculpt the reality we perceive.

The story of the universe, as narrated by humans, has been punctuated by spectacular breakthroughs: radical upheavals that overturned our worldview while diminishing our perception of our place in it. From the time we declared the Earth as the central, unmovable stage around which the celestial drama unfolded, we've revised the story to take in the staggering scale of the universe and our truly diminutive presence within it. Having adjusted to our new, spatially peripheral perspective, we then asked: What about life? How common is life in the cosmos? Although life on Earth had to navigate exceptional obstacles, it is plausible – given the extent of the cosmic expanse and the ubiquity of the building blocks – that life is a phenomenon not just confined to our planet but potentially widespread across the cosmos. Our spirited search for life elsewhere may yet yield a breakthrough that shatters our place in the biological order of the universe.

Until such time, how might the future trajectory of humanity play out? How might we expand our role within the cosmos? All the creatures that have ever emerged on our planet have lived and died here; billions of myriad life forms and the countless generations of humans that preceded us have breathed their last on this world of ours. But for the first time in history we're beginning to seriously contemplate a future where those born here are not destined to die here, where the only planet we've ever called home may not be the only world we will inhabit, where future generations might open their eyes in a world very different from this blue marbled sphere.

If we are to become a multi-planetary species, if we are to colonise and expand across the galaxy, then we must also navigate the existential challenges that imperil any advanced civilisation once it acquires the technology to bring about its own destruction. From a cosmological perspective, it may be the height of hubris to think that the story of life and intelligence on Earth reaches its apogee with us. We may consider ourselves to be the pinnacle of terrestrial intelligence, but we do not have a monopoly on our planet's future. If we take a broad enough perspective, we may be here to facilitate an important transition in the future of consciousness, to free it from the constraints that have so far confined it to a limited lifespan and the fate of this planet.

THE COSMIC PERSPECTIVE

To gain that cosmic perspective, we have to zoom out to unfathomably remote timescales and ask: what destiny lies in store for our Sun, other stars, our galaxy, other galaxies and, indeed, for the universe itself?

Looking out into this deep future, our Sun will exhaust its hydrogen fuel in around 5 billion years, its inner core will contract, while its outer layers swell into a red giant and engulf Mercury, Venus and possibly also Earth. But before that, in around 3.5 billion years, the growing luminosity of the Sun will instigate a runaway greenhouse effect through our atmosphere that will almost certainly sterilise the Earth's biosphere and put a definitive end to all life on our planet. The end of all complex life, however, is anticipated to occur much sooner, around 1–1.5 billion years from now, owing to the increasing luminosity of the Sun. Any life forms that are still around then will need an exit plan: if they haven't established settlements on other worlds, they will face the inevitable.

Though exceedingly unlikely, it is possible that natural cosmic interventions could save our planet from our Sun's eventual fate. For instance, we've run simulations to determine the probability that a passing star system might fling the Earth out of our Sun's hold and avert it from being consumed. But the chances of this happening are roughly 1 in 50,000, and the Earth would then become a rock drifting alone in the void, not tethered to any star. Life as we know it would not survive, and the biosphere would still be doomed. Though entirely dwarfed by the energy we receive from the Sun, our planet does have an internal fuel source from the radioactive decay of unstable nuclei, which may generate enough heat to allow liquid water to exist very deep underground.

Another even more unlikely scenario is that a passing star system, instead of stripping the Earth from the Sun and casting it into a starless exile, might capture it instead, such that our planet then orbits a new star. Given that the most common category of star is much smaller than our Sun, burning their fuel much more slowly, our planet would enjoy a new and significantly extended lease of life. Nonetheless, even if Earth were to find itself in orbit around such a star, the chances of the biosphere remaining intact during the star-switching manoeuvre remain low.

So, barring any deliberate interventions and assuming the natural course of events, complex life still has a billion or so years on our planet.

Looking further into the future of our corner of the cosmos, the Milky Way galaxy is a member of the Local Group, a collection of galaxies that includes our neighbouring Andromeda and several dozen dwarf galaxies. In around 6 billion years, our Milky Way is predicted to collide with Andromeda – a merger that will occur over long timescales and see the galaxies gravitationally tugging on each other, pulling, twisting and stretching each other so that the spiral arms that previously defined the individual galaxies are stripped away, reconfiguring their collective shape. Within the galaxies themselves, though, a cosmic collision of this magnitude is very unlikely to have the cataclysmic consequences one might expect. The spaces separating the stars are so vast that it's doubtful there will be any stellar collisions – just a larger galaxy with many

A NEW ERA

more stars and a night sky with roughly twice the brightness. Galactic collisions often also involve the merging of large clouds of hydrogen gas; gravity compresses these gas clouds and incites intense bursts of star formation.

On a much more distant timescale, if the current expansion of the universe continues, individual galaxies and galaxy clusters will recede further and further from one another until they disappear beyond each other's view. In around 100 billion years, each galaxy cluster will become an island, a little universe of its own. Another 100 trillion years from now, the last stars will stop shining, and all the stars in the Milky Way will have switched off. Looking even further into the future, even matter will disintegrate into its basic constituents. And according to our current cosmological model, the universe will ultimately wind down to a point where nothing at all happens. It is projected to be a cold, dark and desolate place, devoid of all activity and all the interesting structure that fills it now.

We have speculated about the prospects of life persisting through these seismic shifts in the universe. If we adopt the highly generalised view of life as an information processing system, as pioneered by Freeman Dyson, then we can determine how it might manage to survive even when the energy available in the universe becomes relatively scarce. Dyson argued that life would need to find ways to become more efficient and proposed that it may do so by lowering its metabolism or its information processing rate and hence its operating temperature. It would have to assume a state similar to hibernation, possibly with long dormant periods, thereby conserving energy to make the diminished energy resources last longer.

From this deep-time perspective, we turn to the future of life on our planet. How has life on Earth endured, and how might it thrive? What existential risks might we face, and what preventative or mitigating strategies can we implement?

EXISTENTIAL RISKS

At least 99.9 per cent of all species that ever graced our planet are now extinct, including other human species. Traces in the geological record reveal a multitude of mass extinction events, around fifteen in the last 500 million years; of these, five wiped out more than half of all species inhabiting the planet at the time. The forces of nature, orchestrating on the grandest scales, have been implicated in many of these events: impacts from comets, mountain-sized asteroids, and supervolcanic eruptions.

It's possible we owe our existence to one such cosmic intervention some 66 million years ago, thought to be a prime catalyst for the mass extinction that claimed almost three-quarters of the life on Earth. A mountain-sized asteroid is thought to have impacted the Earth with such extraordinary velocity that it unleashed vast plumes of gas and dust, blocking out the Sun, drastically altering the climate and rewriting the inventory of species. Life then opportunistically filled the void created by this catastrophe. The knock-on effects from the cosmic collision proved terminal for the dinosaur dynasty, after 165 million years, and in the ecological niches the dinosaurs left open, our mammalian ancestors began to proliferate and dominate, ushering in the Age of Mammals. Had this asteroid missed, life's trajectory here might have been markedly different, and dinosaurs might still have lorded over the Earth.

Through continuous surveillance of space, we have already identified the sizeable asteroids that might pose an extinction-level threat in the future, and, of those, none seem to be on a

A NEW ERA

collision course with Earth. But there may be smaller sized objects that evade our detection system and lurk below the radar. Although not disastrous on a global scale, they could still cause extensive damage. How do we defend our planet from such impacts? One idea pursued by NASA is to smash a spacecraft into the asteroid to alter its course ever so slightly, so that it misses the Earth. Provided this is done far enough in advance, you only need to change the rock's velocity by a small amount. In 2022, NASA deployed this strategy in a proof of principle test demonstrating our first-ever planetary defence mission – the Double Asteroid Redirection Test (DART).

This remarkable defence demonstration took place 11 million kilometres from Earth, where a small 150-metre-wide asteroid, Dimorphos, orbits the larger 760-metre-wide asteroid Didymos. Designed for this mission, the spacecraft *DART* was sent to the system with the objective of crashing into the smaller asteroid, which it would only distinguish from the larger one fifty minutes before impact. Using sophisticated navigation software, the spacecraft deftly engaged its thrusters, making precise adjustments to its trajectory and ultimately impacting Dimorphos at over 20,000 kilometres an hour. The force of this collision ejected over 1 million kilograms of debris into space, producing a 10,000-kilometre comet-like tail that persisted for months and was captured by the Hubble Space Telescope. The impact from the crashed craft and the recoil from the excavated debris altered the rock's orbit around Didymos by a few minutes daily and reduced its orbital period by thirty-three minutes. Even a nudge of this small magnitude would be enough to evade or mitigate against an Earth-bound impactor, provided that the manoeuvre is performed early enough.

But what if we were to spot an asteroid only a few months

away from impact? In such a scenario, a last-ditch option would be to use a nuclear explosive. We've run simulations to determine what scale of nuclear device we would need – a 100-metre-wide rock could be fragmented using a one-megaton nuclear device, with almost all of its mass being blasted out of Earth's way, provided it is done more than two months ahead of time. It would be the last resort but may be a necessary one if we don't have enough time to deploy the deflection strategy.

The other major natural phenomenon that has had a significant impact is supervolcanism. Major eruptions in the past have been responsible for large-scale demolitions of life. One of the greatest volcanic events in the geological record was the eruption of the Toba volcano in Sumatra, Indonesia, around 75,000 years ago, which spewed fine ash and aerosols into the atmosphere, instigating a chain of effects similar to nuclear winter scenarios. The cloud of debris reflected sunlight back into space, causing regional temperatures to drop by 5–15 degrees Celsius on land, leading to a 'volcanic winter'. The longer-term consequences on the climate probably lasted for decades due to feedback effects, such as ever-growing chunks of ice cover that then reflect sunlight back into space. Though debated, the 'Toba catastrophe theory' suggests that the human population may have undergone a near catastrophic drop in numbers around this time, making it, at least in human terms, perhaps the most cataclysmic event ever to have occurred.

Supervolcanoes present more of a challenge than asteroids, because predicting the timing and magnitude of eruptions remains difficult, and, unlike our 'divert or destroy' strategy, we have no credible means of intercepting them. We can, though, try to mitigate their long-term impacts; for instance,

A NEW ERA

by planning for the disruption of agricultural production, with its knock-on effects of mass starvation and social upheaval, and ensuring significant stockpiles of grain and other foodstuffs.

Since we acquired mastery over the atom and its constituents and learned to weaponise its latent potential, the spectre of nuclear war has loomed large as one of our existential threats. Though, until very recently, receding in the public consciousness since the end of the Cold War, it continues to be a menacing prospect. In the event of a full-scale nuclear confrontation, the projected figures are sobering. But even beyond the immediate devastating consequences is the possibility that detonating many of these devices at once could trigger a nuclear winter, where vast plumes of smoke and soot darken the skies, obstructing sunlight and radically shifting weather patterns. Even a small-scale nuclear conflict could initiate global cooling, though the extent of this remains uncertain and is contingent on the type and volume of aerosols released. Historically, the fear was of total human extinction due to radiation and contamination; however, ongoing research indicates that this is less likely than once thought. Even the most catastrophic nuclear winter scenario, it seems, would not spell the end for humanity.

While impacting asteroids or super powerful volcanic eruptions evoke a David versus Goliath struggle against the forces of nature, humanity has been brought to its knees many times by a far less imposing nemesis: microbes. The microorganisms with which we share our planet are among our oldest adversaries. The Black Death originated in East Asia and swept across Central Asia and into Europe via the medieval Silk Road trade routes, lasting in Europe until the early nineteenth century and claiming an estimated 200 million lives. It decimated up to

30 per cent of Europe's population in the fourteenth century and struck in successive waves in Milan, London and Marseilles in the seventeenth and eighteenth centuries. The Spanish flu of 1918–19 claimed 20–50 million lives, though it was overshadowed in the public psyche by the casualties from the First World War. And yet, we've encountered less than 1 per cent of the vast arsenal of microbes. Our exposure to dangerous pathogens is only likely to increase due to the heightened contact between humans and animals, our encroachment into their territory, the effects of climate change and the resulting loss of habitat, and, of course, our global interconnectedness, which has made it all the more likely for a single pathogen to swiftly saturate our society (as shown by the Covid-19 pandemic).

Developments in biological technology are making it easier to create designer pathogens. We've managed to synthesise the polio virus from scratch and have even resurrected the extinct 1918 Spanish flu virus. Complete genomes of hundreds of viruses and bacteria have been sequenced and are available. With the technological barriers to producing superbugs becoming lower, the likelihood of them being manufactured in small, easily concealed facilities and deployed for nefarious mass-scale destruction is growing. And with more labs handling dangerous pathogens comes the increased risk of accidental releases.

Featuring high on the existential risks register is the gradual warming of the climate. While our planet's climate has naturally cycled between ice ages and interglacial periods roughly every 100,000 years over the past million years, the current warming trend is primarily driven by human activity. To maintain a steady climate, there must be a balance between the energy coming in and the energy going out. It's a system so delicately poised that even minor disruptions to its equilibrium can induce global-scale climate effects, majorly impacting the ecosystems in which life of all kinds thrives. Historically, solar variations and volcanic activity have been the main influencers of climate change. But the key driver now appears to be greenhouse gases based mainly on carbon emissions caused by human activity. Climate modelling is a complex system and, while we've made significant progress in understanding it, future warming is still fraught with a degree of uncertainty. Nonetheless, it is evident that climate change presents a challenge we must address to safeguard our long-term future on this planet.

Given the silence of the stars and the apparent absence of other civilisations, we remain especially cautious that the Great Filter – that event that may have precluded others from contacting us – may lie ahead of us. Hence, being attentive to all threats to our immediate and long-term survival is a prudent strategy. Throughout history, forecasting humanity's future has been an inherently unreliable exercise: just as our ancestors could not have foreseen the world we find ourselves in right now, our current knowledge will probably be dwarfed by future discoveries, tossing aside all our projections and upending our best guesses. We're especially unable to guard against so-called 'Black Swan' events – those rare but extreme events that have a disproportionate impact and could not have been predicted from past data (for instance, the September 11th terrorist attacks). Projections for the future are laden with inherent biases and limitations, carrying all the imprints of our cultural era and the existential risks that have struck a chord in our collective consciousness.

A BRIEF HISTORY OF THE UNIVERSE

Immersed as we are in the digital age, one possibility that has rapidly risen to the forefront of public consciousness is a future in which machines take over.

FUTURE OF INTELLIGENCE

Our unique cognition, with its remarkable capacity for abstraction, has allowed us to translate the complexities of an imposing universe into elegant mathematical equations. Leveraging the collective power of this intelligence has brought us great knowledge of the world we inhabit. But is our species really the pinnacle of complex intelligence on Earth? What if we were to engineer an intelligence, biological or artificial, that was capable of upgrading itself – not constrained by the need to toil long and hard building upon generations of ancient wisdom, but instead able to evolve at an unprecedented pace? Within the span of a single generation, an extraordinary technological revolution has brought what were once the fantastical imaginings of science fiction into the boundaries of our reality.

Sixty years ago, in a remote village nestled among the rolling green hills of north Pakistan, my grandfather came across a newspaper article forecasting a not-so-distant future in which we would not only be able to hear the person on the other end of the telephone, but also be able to see them. My father recalls their collective disbelief that such a futuristic scenario was even a possibility – in those days of the corded, rotary-dial telephones, the notion seemed incredible. Yet, within a few short decades, face-calling technology had been rolled out, and this once fanciful notion had become an everyday occurrence. Now, I sit under a clear night sky in that same village, watching the overhead procession of a distinctive trail of bright lights – the

A NEW ERA

Starlink satellite. From rotary phones to global satellite networks, technology has rapidly and probably irreversibly transformed our lives. But the technologies we wield today – and those just over the horizon – have the power to overhaul not just how we live but who we are, and who or what we will become.

Many think it inevitable that machines will eventually take over. Some predict that human history is fast approaching a 'technological singularity' – a tipping point when rapid technological progress induces such seismic changes in human affairs that life as we know it is fundamentally altered. In physics, a singularity is a point in space and time where our laws break down and we can no longer make any predictions for what happens; these occur, for instance, at the centres of black holes or at the instant of the Big Bang. Similarly, a technological singularity renders our ability to predict what happens beyond it obsolete.

En route to this hypothetical event is the milestone of creating an intelligence that is on par with human intellect and capabilities, not just in a narrow task like playing chess but in a whole range of domains. If this intelligence then becomes capable of upgrading itself, we could be looking at an 'intelligence explosion' – where it undergoes continuous self-improvement and redesigns itself at an accelerating rate, such that it far exceeds our ability to even comprehend it. We would then be in the realm of superintelligence. What our brains have acquired over millions of years of evolutionary selection could be exponentially outstripped in a matter of moments. And the actions and decisions of such an intelligence would be inscrutable to us. In short, it would have the potency to make all our previous technologies seem like child's play.

A BRIEF HISTORY OF THE UNIVERSE

Should we succeed at designing an intelligence that outsmarts our own, we would invariably be heading towards an alternate future for which we have no historical analogue. A superintelligence could rapidly accelerate innovations and, in minutes, solve complex problems that might otherwise take us generations. It's likely to rewrite the rules of almost every aspect of human existence while immersing us in a host of ethical and philosophical dilemmas. How do we govern an entity whose intellect is vastly superior to our own and ensure it acts in humanity's best interests? What moral obligations would we have towards it, and how do we deem whether it qualifies as sentient? The social implications could be equally staggering. Many jobs may be rendered obsolete, liberating us to engage in more leisurely pursuits while, paradoxically, plunging us into existential crises of identity and finding purpose and value without work.

Of the future trajectories that may rewrite our place in the cosmos, this has swiftly become one that merits the most consideration. If we do cross this threshold, will we come to see the era of human-level biological intelligence as a mere prelude to the emergence of superintelligent machines? Unimpeded by the substantive demands of a biological substrate, machine intelligence may be capable of outlasting us by many epochs. Parallel developments in robotics might enable this intelligence to become embodied in a physical form, one that will be much more resilient and versatile in the diverse environments beyond Earth. At the same time, we may augment our own cognitive and physical capabilities, editing our DNA using tools like CRISPR-Cas9, enhancing our intellectual faculties and physical capabilities, and even extending our lifespans.

A NEW ERA

An intimate merging of humans and machines may be how we overcome the attendant limitations of our biology. The more we learn about how the brain works and the parts of it responsible for various aspects of our cognition, the greater our ability to tinker with it. Major developments are driving the field of brain–computer interfaces, which create a direct communication pathway between the brain's intricate neural networks and external devices. The current technology is being employed to enable those with paralysis to control computer cursors, robotic arms and even their own limb prosthetics through thought. Implantable versions of these interfaces have allowed those with spinal cord injuries to regain control over their environment.

On the spectrum of possibilities for our integration with machines, perhaps the most radical is 'mind uploading', where a mind is migrated onto a machine. This would be like attaining a form of 'digital immortality' where we discard our 'biological shell' and assume a digital existence on a substrate that does not suffer from the trappings of our organic form. Of course, it remains highly speculative, and even if we were to engineer the technology, it raises a famously difficult and important philosophical question about the nature of consciousness. How does a physical entity create the rich subjective experience we associate with being conscious, and can it really be captured and replicated on an alternative computational basis? How can we even tell that an uploaded mind is conscious and not merely 'acting' the way a human would? Does the inherent subjectivity of consciousness make it impossible to assert with absolute confidence that another entity is conscious? There's much debate on both the technological feasibility of mind uploading and its philosophical implications. Some see it as being our only option for travelling to the stars, sending 'e-crews' or mind-uploaded astronauts that circumvent the limitations of biological humans, given the vast distances involved.

It is beyond doubt that machines will have a prominent role in our unfolding future. Will we be entirely usurped by them and fade into the oblivion of the animal kingdom? Will we try to keep up by augmenting our physical and mental capabilities? Or is an intimate merger of humans and machines our destiny? Having glimpsed the immense potential of intelligent machines, will we be able to – or even want to – resist taking a bite from the apple? As with any supremely powerful technology, the more possibilities there are for creating utopias, the more dystopian the potential risks. Are we ready for this step change in our existence? Have we the wisdom to develop technologies that may undermine our identity, our legacy, our future and potentially the future of all life on earth? There are voices calling for a staggering of this process, at the very least, so that we may fully understand the implications and where they could lead. We don't know which of the possible futures we will find ourselves in, but given the stakes, exercising restraint so we can appreciate the repercussions seems wise. Time will tell if *Homo sapiens* lives up to the name it bestowed on itself – or whether future generations declare this our greatest irony.

CHAPTER 13

OUR COSMIC LEGACY

Will the next transformative leap in our history see us shedding our biological constraints to assume a digital existence in the cosmos? Will the essence of *Homo sapiens* be carried forth not in its carbon incarnation, but in a digital avatar that bears the hallmarks of its creator, unburdened by its biological limitations? If we are at the precipice of changes that may affect Earth and all its inhabitants, it is worth asking what kind of legacy we wish to leave in the cosmos. Do we have a responsibility to preserve life and even ensure its continuity beyond our planet? Even if life is abundant in the cosmos, it's exceedingly unlikely that the expression it has found on Earth has been replicated anywhere else. The magnificent blue whale that graces our oceans is very probably only one of its kind in the entire universe, as is the bustling column of ants that scurry about for the good of their colony, and so too goes for all of us.

A BRIEF HISTORY OF THE UNIVERSE

ATTUNED TO EARTH

For the next billion or so years, complex life has a future on Earth, and it is far better tuned to this planet than to any other. We might scour the depths of the cosmos and not find another world more agreeable to us than this.

In cosmic time, the emergence of our species, our divergence from the great apes, our acquisition of language and symbolic thinking, our transmission of ideas through cultural evolution, our raising of great civilisations, our mastery of technology and our enquiry of the cosmos have all unfolded in what amounts to a mere blip. But the seeds that led to our emergence were sown billions of years ago, shortly after the Earth formed. We, along with all other species on this planet, have co-evolved with it and are deeply and inseparably attuned to this Goldilocks-zone blue rock. Our sense perceptions, bodily structure and psychological make-up have been tuned and tailored to thrive within the parameters of this planet's environment.

Adapted to the spectrum of sunlight filtered through the Earth's atmosphere, our eyes see the world through a vivid prism of colours. Specialised retinal cells enrich us with trichromatic vision, an adaptation that may have helped our ancestors find ripe fruit in dense foliage and discern each other's changing moods through subtle flushes in skin. Our forward-facing binocular vision provides the depth perception that once enabled our tree-dwelling ancestors to judge distances with precision, while our pupils continually adjust their circumference to adapt to varying light conditions.

Our skin is sensitive to the blistering heat of our star and has evolved mechanisms to protect and regulate our internal environment; sweat glands cool us in the heat, and constricting blood vessels retain heat in the cold, while melanin shields us from harmful UV radiation, acting as a natural sunscreen and varying across populations to adjust to the amount of sunlight they typically encounter.

Our digestive system has been precisely calibrated to harness the bounty of Earth's flora and fauna. Enzymes dismantle the complex molecules found in our planet's provisions, from the starch stored in plant cells to the proteins that scaffold animal tissues. The trillions of microorganisms that line our gut have also co-evolved with us, and we rely on this army of bacteria to ferment dietary fibres that our own enzymes cannot digest, combat pathogenic bacteria, modulate our immune system and even shore up our mental wellbeing.

Our internal clock, along with that of plants, animals and microorganisms, is synchronised with our planet's daily rotation. Tuned to Earth's twenty-four-hour cycle of day and night, our circadian rhythm directs a roster of bodily functions, energising us as the sun rises and winding us down when daylight recedes.

Our bones and muscles have grown strong under the influence of Earth's gravity. The architecture of our bodies, their skeletal and muscular structure, has adapted to the demands of moving and operating within the constraints of our planet's pull. Our bone density and muscle strength support our vertical, bipedal posture, with which we navigate and inhabit Earth's diverse terrains, from the plains of the African savannah to the rugged hilltops of the Himalayas.

Should we travel to the stars and set up outposts on other celestial bodies, we will have to equip our biology to withstand environments utterly alien to our evolutionary adaptations. In the absence of Earth's gravitational field, the microgravity of space will challenge our muscular and skeletal integrity, a phenomenon astronauts counter with rigorous exercise in orbit. The cosmos is awash with radiation, from highly energetic galactic cosmic rays that can tear through cellular material to the sporadic outbursts from our Sun. Cloaked beneath Earth's magnetic field and atmospheric shield, we are spared the full force of this cosmic assault. But in space, without this protection, we'd be exposed to levels of radiation that can severely damage our DNA, rendering long-term space habitation and interstellar voyages particularly perilous.

Our psyches have developed in step with the demands of our social structures. We are not solitary creatures who can endure long periods of isolation, but intensely social beings; our mental faculties honed over millennia to navigate the dynamics of complex social organisation. We respond poorly to loneliness. Deprived of contact with others, we conjure the social world that we're lacking in reality, anthropomorphising our pets, become obsessively attached to fictional characters on screen, and lingering in the sentimental nostalgia of past social interactions.

Pioneering voyages across the cosmos will entail long periods being stationed in an unfamiliar environment, accompanied by only a handful of others. Even on short missions, astronauts undergo rigorous psychological training to mentally brace for the exile and confinement of space travel. And, despite this, many feel a heightened sense of solitude and vulnerability when confronted with the vastness of space. For the psyche that has developed to be surrounded by others, space is about as unnatural an environment as there can be.

The machine intelligence we've created does not require air, water or warmth, nor the physical embrace of others or their social companionship. It can potentially withstand much more hostile environments than our frail bodies and would require far fewer modifications to meet the demands of space and other planetary environments. Fewer obstacles stand in its way than those we must overcome to transcend the limitations of our Earth-evolved biology. Machines are far more equipped to meet the demands of space exploration, and far less vulnerable in the face of it.

On our behalf and directed from Earth, our machines have explored more of the cosmos than we have. Our Moon has been visited by many orbiters and robotic landers. Neighbouring Mars is being charted by a fleet of rovers scouring its ruddy terrain while orbiters map its surface from above. Inhospitable Venus, shrouded in dense, toxic clouds and subjected to infernal temperature and crushing pressures, has been visited several times. Mercury, with its thin atmosphere, extreme temperatures and intense bombardment by violent solar winds, has been circled by an orbiting craft. Mighty gas giant Jupiter has been photographed by the spacecraft *Galileo*, and now *Juno*. Through the lens of *Cassini*, we've witnessed in spectacular detail the splendour of Saturn's rings and its wondrous, potentially habitable moons. Ice giants Uranus and Neptune were glimpsed by the *Voyager 2* probe as it hurtled towards the edge of our solar system. Our machines have even rendezvoused with the silent wanderers of the cosmos – asteroids and comets – touching down and extracting samples that reveal the ancient secrets of our stellar neighbourhood. Emissaries from Earth have already spread out across the solar system, and our *Voyager* missions have even ventured into the space between the stars.

Yet, so far, these machines serve us much as the primitive

tools our ancestors wielded to conquer their earthly domains. They have been obliging instruments in our exploration of the cosmos. As impressive as their excursions have been, they remain extensions of our will, dispatched on our behalf and directed by our scientific counsel. We cannot be so sure of the technologies that lie ahead, for they promise to be radically more powerful, more autonomous and more intelligent than the minds that designed them. We do not know if we will be able to retain control over them or if they will be guided by their own emergent whims and wishes. We cannot even say what their objectives might be, because speculating on the intentions and goals of a higher intelligence is beyond our purview. That the future of intelligence on Earth may not end with us, that new forms of life and consciousness may take our place, are among the possible futures we must contend with. And these new entities we've given birth to may be far more capable of comprehending the cosmos and expanding into it.

But will they feel the same rapture at the sight of the stars? Will the enormity of the universe stir in them the same intensity of emotions? Will they yearn to scale the heights that separate them from the heavens? Will they experience the sublime ecstasy of being able to comprehend a universe so vast and intricate and essential for our existence – yet unaware of our presence? We don't know how the deep future will unfold, but on the near horizon we should at least ensure that we, and the intelligence and consciousness we embody, have a seat at the table.

EPILOGUE

From the earliest days of our awakening, the rapturous wonder of the world in which we find ourselves has held our collective fascination. Since then, we have striven to understand who we are and how we fit into the grander scheme. Every culture has sought for itself a deeper meaning, a higher purpose, a desire for an eternal truth, a way to connect with an ultimate reality. This search for meaning has not dissipated with our greater understanding of the world. It has just taken on new forms and new philosophies and seeks new narratives in a world so rapidly changing that the scared and vulnerable ape residing deep inside us is forgotten. We were not born in the Age of Machines, nor are we the sole products of the Age of Reason. Our existence has been sculpted over billions of years of evolution, and we carry that extensive lineage in the deep recesses of our minds and the depths of our emotions.

We still seek meaning, even though reason tells us it may be a futile quest. We have been the beneficiaries of chance events that were the dispassionate outcome of evolution, lacking any plan or purpose. Our existence was highly improbable but not inevitable. Yet, we yearn to tether the tale of it to a greater, grander narrative. We have an inherent need to belong, to be part of a history much richer and more enduring than ourselves. After all, it was thanks to ancient myths and mythologies, to the socially binding forces of great religions and ideologies, that we became a civilisation capable of feats so remarkable that they would have been utterly impossible for a single person to accomplish alone.

Our ancestors once held in reverential awe the lights in the night sky, spinning tales of myth and lore in an effort to assert greater control over the capricious outbursts of nature, to pacify the inane horrors of a world that terrified them, to discover meaning in a world they were soon destined to leave, and to carve out a place for their fleeting existence in the infinity that stretched before them, and the eternity that preceded them and would supersede them.

We now know that the stars are not gods; they are not the messengers of great deities, nor their expressions of anger or wrath or desire; they do not spell the death of kings or the fates of our unborn children; they are distant suns, massive fiery balls of gas careering across the void and piercing our darkness from an unfathomable remoteness; and they have worlds of their own, maybe worlds that rely on their radiant warmth to create forms unbeknown to us, maybe worlds that greet their glorious rising with the ecstasy of birdsong and their setting with the poetry of sonnets.

Even though pushing the limits of rational enquiry has brought us great knowledge of the world we inhabit, and mastery over it to a degree we once could not have imagined, we remain a profoundly philosophical species. We are story-

EPILOGUE

tellers, in search of the greatest story of them all, even if we may be a mere footnote in it. Our exploration of the universe is not just a march at the frontiers of our science, but also the soul-seeking venture of a species riddled with impossible questions. Having long been seduced by the lights in our night skies, we're destined for the very stars we once worshipped as gods. What remains to be seen is *how* we get there.

We may hope that amid these sea changes, we do not forget who we are and where we came from: the billions of years of evolution that gave rise to our own genesis, the innumerable creatures with whom we share an intense fellowship, and the planet that has nourished our growth and surrendered its resources for our adventurous pursuits. As we venture into an unknown future we remind ourselves that we remain a vulnerable species in search of meaning, and facing a brave new world.

ACKNOWLEDGEMENTS

I'm grateful to my agent, Kate Evans, for her endless enthusiasm; Assallah Tahir for seeing the vision; Kate Harvey for her meticulous editorial work; Allie Johnston for her valuable feedback; and the entire team at Simon & Schuster UK for their consistent support.

Thanks to Natalie James for encouraging me to do this in the first place, and to Callum Duffy and Marcin Jastrzebski for sense-checking some paragraphs. A special thanks to Zackaria for reading through the entire manuscript – which alone has made this endeavour worthwhile.

I also remain deeply indebted to my family – in particular, my parents, siblings, parents-in-law, and most of all, Dawood.

BIBLIOGRAPHY

Chapter 1: The Ancient Cosmos

Alhazen (Ibn al-Haytham), *Shukūk 'alā Baṭlamyūs* (Doubts on Ptolemy). Eleventh century.

Jim Al-Khalili, *The House of Wisdom: How Arabic Science Saved Ancient Knowledge and Gave Us the Renaissance* (London: Penguin Books, 2011).

Jonathan Barnes, *Aristotle: A Very Short Introduction*. Very Short Introductions (Oxford: Oxford University Press, 2000; online edn, Oxford Academic, 26 Nov. 2015), https://doi.org/10.1093/actrade/9780192854087.001.0001

T. Freeth, D. Higgon, A. Dacanalis et al., 'A Model of the Cosmos in the ancient Greek Antikythera Mechanism', *Sci Rep* 11, 5821 (2021), https://doi.org/10.1038/s41598-021-84310-w

Michael Hoskin, 'Astronomy in antiquity', *The History of Astronomy: A Very Short Introduction*. Very Short Introductions (Oxford: Oxford University Press, 2003; online edn, Oxford Academic, 24 Sept. 2013), https://doi.org/10.1093/actrade/9780192803061.003.0002

https://ocw.mit.edu/courses/sts-003-the-rise-of-modern-science-fall-2010/resources/mitsts_003f10_lec17/

Ptolemy, *Almagest*, trans. G. J. Toomer (Princeton, NJ: Princeton University Press, 1998).

A. I. Sabra, ed., *The Optics of Ibn al-Haytham*, Books I–II–III: On Direct Vision. English translation and commentary, 2 vols. Studies of the Warburg Institute 40, trans. A. I. Sabra (London: The Warburg Institute, University of London, 1989).

Abraham J. Sachs and Hermann Hunger, *Astronomical Diaries and Related Texts from Babylonia*, 3 vols (Vienna: Verlag der Österreichischen Akademie der Wissenschaften, 1988–1996).

George Saliba, *A History of Arabic Astronomy: Planetary Theories During the Golden Age of Islam* (New York: New York University Press, 1994).

Katharina Volk, 'Chapter 2 Portrait of the Universe', in *Manilius and His Intellectual Background* (Oxford: Oxford University Press, 2009), pp. 31–68. Accessed 1 May 2009. Oxford Academic.

Chapter 2: The Copernican Revolution

Karen Armstrong, *A History of God* (New York: Alfred A. Knopf, 1993).

W. Bogdanowicz, M. Allen, W. Branicki, M. Lembring, M. Gajewska and T. Kupiec, 'Genetic identification of putative remains of the famous astronomer Nicolaus Copernicus', *Proc. Natl. Acad. Sci. U.S.A.* 106 (30), 12279–82, (2009) ttps://doi.org/10.1073/pnas.0901848106

Galileo Galilei, *Discoveries and Opinions of Galileo*, trans. Stillman Drake (Garden City, NY: Doubleday Anchor Books, 1957).

Galileo Galilei, *Sidereus Nuncius, or The Sidereal Messenger*, trans.

Albert Van Helden (Chicago: University of Chicago Press, 1989).

Owen Gingerich, *Copernicus: A Very Short Introduction*. Very Short Introductions (Oxford: Oxford University Press, 2005). Owen Gingerich, *The Eye of Heaven: Ptolemy, Copernicus, Kepler* (New York: American Institute of Physics, 1993).

Thomas L. Heath, *Aristarchus of Samos: The Ancient Copernicus*. Reprint (Mineola, NY: Dover, 2004).

Chapter 3: Pushing Boundaries

J. C. Adams, 'Explanation of the Observed Irregularities in the Motion of Uranus, on the Hypothesis of Disturbance by a More Distant Planet; with a Determination of the Mass, Orbit, and Position of the Disturbing Body', *Monthly Notices of the Royal Astronomical Society* 7 (1846), pp. 149–52, https://doi.org/10.1093/mnras/7.9.149.

David Brewster, *Memoirs of the Life: Writings, and Discoveries of Sir Isaac Newton*, 2 vols (Edinburgh: Thomas Constable and Co., 1855).

C. A. Chant, 'Johann Gottfried Galle', *Journal of the Royal Astronomical Society of Canada*, vol. 4 (1910), p. 379.

Louis A. Girifalco, 'The seeker', *The Universal Force: Gravity – Creator of Worlds* (Oxford, 2007; online edn, Oxford Academic, 1 Jan. 2008), https://doi.org/10.1093/acprof:oso/9780199228966.003.0001

Edmond Halley, *A synopsis of the astronomy of comets* (printed for John Senex, 1705), doi: https://doi.org/10.5479/sil.271675.39088015653660

Rob Iliffe, *Newton: A Very Short Introduction*. Very Short Introductions (Oxford: Oxford University Press, 2007. Online edition, Oxford

Academic, 24 Sept. 2013), https://doi.org/10.1093/actrade/9780199298037.001.0001.

C. Kowal and S. Drake, 'Galileo's observations of Neptune', *Nature* 287 (1980), pp. 311–313, https://doi.org/10.1038/287311a0

Urbain Jean Joseph Le Verrier, 1846. 'Sur la planète qui produit les anomalies observées dans le mouvement d'Uranus', *Comptes rendus* 23, pp. 428–38, doi:10.1002/asna.18460230303

Simon Schaffer, 'Uranus and the Establishment of Herschel's Astronomy', *Journal for the History of Astronomy* 12, no. 1 (1981).

Matthew Turner, in 'The Theological and Miscellaneous Works of Joseph Priestley', J. T. Rutt, ed., vol. 1, London, 1817), pp. 76.

Chapters 4 and 5: Not Like Clockwork

Timothy Clifton, *Gravity: A Very Short Introduction*. Very Short Introductions (Oxford: Oxford University Press, 2017).

Olivier Darrigol, *Relativity: Principles and Theories from Galileo to Einstein* (New York: Belin, 2005).

Bryan Magee, *Confessions of a Philosopher: A Personal Journey Through Western Thought* (New Haven: Yale University Press, 1997).

John Stachel, *Einstein from 'B' to 'Z'* (Boston: Birkhäuser, 2002).

Russell Stannard, *Relativity: A Very Short Introduction*. Very Short Introductions (Oxford: Oxford University Press, 2008).

Chapters 6 and 7: Building Blocks

Carl D. Anderson, https://calteches.library.caltech.edu/526/2/Anderson.pdf

Frank Close, *'How empty is an atom?'*, *Nothing: A Very Short Introduction*. Very Short Introductions (Oxford: Oxford University Press,

BIBLIOGRAPHY

2009; online edn, Oxford Academic, 24 Sept. 2013), https://doi.org/10.1093/actrade/9780199225866.003.0002

Frank Close, *Particle Physics: A Very Short Introduction*. Very Short Introductions (Oxford: Oxford University Press, 2023).

Geoff Cottrell, *Matter: A Very Short Introduction*. Very Short Introductions (Oxford: Oxford University Press, 2019).

Samir Okasha, *Philosophy of Science: A Very Short Introduction*. Very Short Introductions (Oxford: Oxford University Press, 2002).

Ernest Rutherford, 'The Scattering of α and β Particles by Matter and the Structure of the Atom', *Philosophical Magazine* 6, no. 21 (1911), pp. 669–88.

2018 documentary movie *Salam: the First ****** Nobel Laureate*. http://kailoola.com/salam/

Eric Scerri, *The Periodic Table: Its Story and Its Significance* (New York, 2019; online edn, Oxford Academic, 12 Nov. 2020), https://doi.org/10.1093/oso/9780190914363.003.0012

H. K. M. Tanaka, K. Sumiya, and L. Oláh, 'Muography as a new tool to study the historic earthquakes recorded in ancient burial mounds', *Geosci. Instrum. Method. Data Syst.* 9, 357–364, https://doi.org/10.5194/gi-9-357-2020, 2020.

G. P. Thomson, 'JUBILEE OF THE DISCOVERY OF THE ELECTRON', *Nature* 160 (1947), no. 4058, p. 176.

Chapter 8: Dark Universe

R. A. Alpher, H. Bethe and G. Gamow, 'The Origin of Chemical Elements', *Physical Review* 73 (1948), pp. 803–4, https://doi.org/10.1103/PhysRev.73.803

R. A. Alpher and R. Herman, 'Evolution of the Universe', *Nature* 162 (November 1948), pp. 774–5, https://doi.org/10.1038/162774b0

R. H. Dicke, P. J. E. Peebles, P. G. Roll and D. T. Wilkinson, 'Cosmic Black-Body Radiation', *Astrophysical Journal* 142 (1965), pp. 414–19.

Kathyn Jepsen, 'Vera Rubin: Giant of Astronomy', *Symmetry Magazine*, https://www.symmetrymagazine.org/article/vera-rubin-giant-of-astronomy

John Johnson, Jr, *Zwicky: The Outcast Genius Who Unmasked the Universe* (Washington DC: National Academies Press, 2019).

Carl Lankowski, Pam Lankowski and Vera Rubin, 'Oral Histories', Historic Chevy Chase DC, 05 November 2011. https://www.historicchevychasedc.org/oral-histories/vera-rubin/#:~:text=do%20and%20I%20said%20I,Believe%20me%2C%20I%20felt

Arno A. Penzias and Robert W. Wilson, 'Measurement of Excess Antenna Temperature at 4080 Mc/s', *Astrophysical Journal* 142 (1965), pp. 419–21.

Vera C. Rubin, and W. Kent Ford Jr, 'Rotation of the Andromeda Nebula from a Spectroscopic Survey of Emission Regions', *Astrophysical Journal* 159 (1970), pp. 379–403.

Walter Sullivan, 'Signals Imply a "Big Bang" Universe', *New York Times*, 21 May 1965. https://www.nytimes.com/1965/05/21/archives/signals-imply-a-big-bang-universe-signals-imply-a-big-bang-universe.html

David Wallace, *Philosophy of Physics: A Very Short Introduction*. Very Short Introductions (Oxford: Oxford University Press, 2021).

R. W. Wilson, History of the Discovery of the Cosmic Microwave Background Radiation, *Physica Scripta*, 21, no. 5 (1980), pp. 599.

Chapter 9: Life As We Know It

S. F. Cabreira, C . L. Schultz, L. R. da Silva, L. H. P. Lora, C. Pakulski, R. C. B do Rêgo et al., 'Diphyodont tooth replacement of

BIBLIOGRAPHY

Brasilodon – A Late Triassic eucynodont that challenges the time of origin of mammals', *Journal of Anatomy*, vol. 241,6 1–17 (2022). Available from: https://doi.org/10.1111/joa.13756

A. S. Eddington, *The Nature of the Physical World* (1928).

J. L. England, 'Dissipative adaptation in driven self-assembly', *Nature Nanotechnoly* 10 (11), 919-23 (November 2015). doi: 10.1038/nnano.2015.250. PMID: 26530021.

J. L. England, *Every Life Is On Fire: How Thermodynamics Explains the Origins of Living Things* (New York: Basic Books, 2020).

Peter Goddard, *Other Minds: The Octopus, the Sea, and the Deep Origins of Consciousness* (New York: Farrar, Straus and Giroux, 2016).

Lucretius, *On the Nature of Things*, trans. Anthony M. Esolen (Baltimore: Johns Hopkins University Press, 1995).

E. R. R. Moody, S. Álvarez-Carretero, T. A. Mahendrarajah et al., 'The nature of the last universal common ancestor and its impact on the early Earth system' *Nat Ecol Evol* 8, 1654–1666 (2024), https://doi.org/10.1038/s41559-024-02461-1

Erwin Schrödinger, *What Is Life?: The Physical Aspect of the Living Cell* (Cambridge: Cambridge University Press, 1944).

John Maynard Smith and Eors Szathmary, *The Major Transitions in Evolution* (Oxford: Oxford University Press, 1997).

Chapter 10: Life in Our Neighbourhood and Beyond

Jim Al-Khalili, ed., *Aliens* (London: Bantam Press, 2023).

Nathalie Cabrol, *The Secret Life of the Universe: Exploring the Cosmic Origins of Life* (New York: St Martin's Press, 2023).

https://science.nasa.gov/resource/solar-system-temperatures/

https://gulfnews.com/uae/government/uae-unveils-details-of-uae-mars-mission-1.1505710

https://www.mmx.jaxa.jp/en/

https://www.isro.gov.in/MarsOrbiterMissionSpacecraft.html
https://www.emiratesmarsmission.ae
https://science.nasa.gov/mission/voyager/interstellar-mission/
https://lasp.colorado.edu/mop/files/2018/07/Chapter-2.pdf
https://www.space.com/42848-earthrise-photo-apollo-8-legacy-bill-anders.html
https://www.theatlantic.com/technology/archive/2017/06/solving-the-mystery-of-whose-laughter-is-on-the-golden-record/532197/
https://science.nasa.gov/mission/kepler/
https://www.isas.jaxa.jp/en/missions/spacecraft/past/ikaros.html
https://breakthroughinitiatives.org

Kepler and its extended mission, K2, surveyed half a million stars.

Robert B. Leighton, Bruce C. Murray, Robert P. Sharp, J. Denton Allen and Richard K. Sloan, 'Mariner IV Photography of Mars: Initial Results', *Science* 149, no. 3684 (6 August 1965), pp. 627–30.

R. J. McKim, 'Astronomy on Mars Hill', *Journal of the British Astronomical Association* 105 (1995), pp. 69–74.

Quote from Ruler of United Arab Emirates Shaikh Mohammad bin Rashid: 'Arab civilisation once played a great role in contributing to human knowledge and will play that role again.'

Robert Poole, *Earthrise: How Man First Saw the Earth* (New Haven: Yale University Press, 2008).

'Schiaparelli's Observations of Mars', *The Observatory* 5 (1882), pp. 138–143.

'Solar Radiation and the Earth's Energy Balance'. The Climate System – EESC 2100 Spring 2007. Columbia University.

The *Cassini–Huygens* mission is a cooperative project of NASA, ESA and the Italian Space Agency (ASI).

A. Wolszczan and D. Frail, 'A planetary system around the millisecond pulsar PSR1257 + 12', *Nature* 355, 145–147 (1992), https://doi.org/10.1038/355145a0

BIBLIOGRAPHY

Chapter 11: Decoding Messages from the Cosmos

J. A. Ball, 'The Zoo Hypothesis', *Icarus* 19, no. 3, pp. 347–9 (1973). doi:10.1016/0019-1035(73)90111-5

John D. Barrow, *Impossibility: The Limits of Science and the Science of Limits* (New York: Oxford University Press, 1998).

Davide Castelvecchi, 'Universe Has Ten Times More Galaxies than Researchers Thought', *Nature*, 14 October 2016, https://doi.org/10.1038/nature.2016.20809 Freeman J. Dyson, 'Search for Artificial Stellar Sources of Infrared Radiation', *Science* 131, 1667–8 (1960). DOI:10.1126/science.131.3414.1667, https://www.nasa.gov/missions/kepler/about-half-of-sun-like-stars-could-host-rocky-potentially-habitable-planets/

A. S. Eddington, *The Nature of the Physical World* (1928).

https://www.nasa.gov/missions/kepler/about-half-of-sun-like-stars-could-host-rocky-potentially-habitable-planets/

https://www.seti.org/protocols-eti-signal-detection-0

https://www.sonarcalling.com/en/

https://astronomynow.com/news/n1004/26seti5/

Nikolai S. Kardashev, 'Transmission of Information by Extraterrestrial Civilizations', Soviet Astronomy, vol. 8 (1964), p. 217.

Abel Méndez, Kevin Ortiz Ceballos and Jorge I. Zuluaga, 'Arecibo Wow! I: An Astrophysical Explanation for the Wow! Signal', arXiv:2408.08513v2 [astro-ph.HE], 2024 National Telecommunications and Information Administration, Federal Government Spectrum Compendium: 1400.00-1427.00, 1 March 2014

Stephen Hawking's Favorite Places. Directed by Stephen Mizelas, 2016. Streaming video. CuriosityStream https://app.curiositystream.com/video/1697

Nick Tusay, Sofia Z. Sheikh, Evan L. Sneed, Wael Farah, Alexander

W. Pollak, Luigi F. Cruz, Andrew Siemion, David R. DeBoer and Jason T. Wright, 'A Radio Technosignature Search of TRAPPIST-1 with the Allen Telescope Array', *Astronomical Journal* 168, no. 6 (2024), p. 283.

Alexander Zaitsev, 'Messaging to Extra-Terrestrial Intelligence', arXiv preprint [physics/0610031], 3 Oct. 2006.

Chapter 12: A New Era

F. C. Adams and G. Laughlin, *Five Ages of the Universe: Inside the Physics of Eternity* (New York: The Free Press, 1999).

Nick Bostrom and Milan M. Ćirković, eds, *Global Catastrophic Risks* (Oxford: Oxford University Press, 2008).

Ken Caldeira and James F. Kasting, 'The life span of the biosphere revisited', *Nature* 360, no. 6406 (1992), pp. 721–3.

F. J. Dyson, 'Time without end: physics and biology in an open universe', *Rev. Mod. Phys.* 51 (1979), pp. 447–60.

https://dart.jhuapl.edu

https://www.newscientist.com/article/2340837-photo-shows-10000-km-debris-tail-caused-by-dart-asteroid-smash/

James F. Kasting, 'Runaway and Moist Greenhouse Atmospheres and the Evolution of Earth and Venus', *Icarus* 74, no. 3 (June 1988), pp. 472–94, https://doi.org/10.1016/0019-1035(88)90116-9

Patrick K. King, Megan Bruck Syal, David S. P. Dearborn, Robert Managan, J. Michael Owen and Cody Raskin, 'Late-Time Small Body Disruptions for Planetary Defense', *Acta Astronautica* 188 (2021), pp. 367–86, https://doi.org/10.1016/j.actaastro.2021.07.034

D. G. Korycansky, G. Laughlin and F. C. Adams (2001), 'Astronomical Engineering: a strategy for modifying planetary orbits', *Astrophys. Space Sci.*, 275, pp. 349–66.

BIBLIOGRAPHY

Gregory Laughlin and Fred C. Adams, 'The frozen Earth: Binary scattering events and the fate of the solar system', *Icarus* 145, no. 2 (2000), pp. 614–27.

P. J. E. Peebles, Orbits of the nearby galaxies, *Astrophys. J.* 429 (1994), pp. 43–65. Jocelyne Piret and Guy Boivin, 'Pandemics Throughout History', *Frontiers in Microbiology* 11 (2021), https://doi.org/10.3389/fmicb.2020.631736

Planetary Science Journal, 5(11), 255. [Authors suggest that complex land plants may survive for longer than previously estimated, up to 1.86 billion years.]

Martin Rees, *On the Future: Prospects for Humanity* (Princeton, NJ: Princeton University Press, 2018).

Murray Shanahan, *The Technological Singularity* (Cambridge, MA: MIT Press, 2015). Substantial Extension of the Lifetime of the Terrestrial Biosphere.

Mustafa Suleyman, *The Coming Wave: Technology, Power, and the Twenty-First Century's Greatest Dilemma* (New York: Crown, 2023).

Chapter 13: Our Cosmic Legacy

Ebrahim Afshinnekoo et al., 'Fundamental Biological Features of Spaceflight: Advancing the Field to Enable Deep-Space Exploration', *Cell*, vol. 184, 24 (2021), p. 6002. doi:10.1016/j.cell.2021.11.008

https://messenger.jhuapl.edu
https://www.jpl.nasa.gov/missions/juno/
https://science.nasa.gov/mission/galileo/https://www.isas.jaxa.jp/en/missions/spacecraft/current/hayabusa2.htmlhttps://science.nasa.gov/mission/osiris-rex/

INDEX

Abbott, Edwin A.: *Flatland* 106
Academy of Sciences, Paris 27
Adams, John Couch 35
aether 9, 44, 51, 75
al-Amal (probe) 158
Al-Battani 13, 14
alchemy 75, 139
Alexander the Great 9
algebra 13
Alhazen (Ibn al-Haytham): *Kitab al-Manazir* (Optics) 13–14; *Shukūk 'alā Baṭlamyūs* (Doubts on Ptolemy) 14
Allen Telescope Array 191
Alpha Magnetic Spectrometer (AMS-02) 103
alpha particles 78–9
Alpher, Ralph 120
Al-Sufi: *Book of Fixed Stars* 38–9
al-'Ijliyyah (Mariam al-Astrulabi) 13
Al-Tusi 14
Alvarez, Luis 96–7
Anders, Bill 163
Anderson, Carl 89–90, 96
Andromeda (galaxy) 38–9, 114–15, 163, 179, 202
Antikythera 11
antimatter 89, 90, 95, 101–4
Apollo missions: Apollo 8 162–3; Apollo 11 153–4; Apollo 15 58
Aquinas, Thomas 24, 75

Arecibo Radio Observatory, Puerto Rico 68, 172, 193–4
Aristarchus of Samos 19
Aristotle 24, 29–30, 74, 75, 127, 131–2; four-element model 9, 74–6; geo-centrism 8–9, 10, 15, 17–20, 23, 25, 32, 132; spherical Earth, observes 8–9
asteroids 144–5, 180–82, 204–7, 219
astrolabe 12, 13
astrology 4–5, 159
Astronomical Diaries, Babylonian 4
atom: atomic clock 54; defined 76; discovery of/atomic model 73–80, 108, 132, 207; life and 137; nucleus of 66, 69, 79; origins of 121; quantum mechanics and 80–81; subatomic world *see* subatomic world

Babylonians 4–7, 8, 9, 10, 12, 63
Barrow, John 186
Besso, Michele 51–2
Big Bang 98, 99, 101–2, 120–21, 126, 176, 211
Big Ear radio telescope, Ohio 190
Big Freeze 126, 135
Big Rip 127
biological technology 208–9
biosignatures 174
Bjorken, James 91
Black Death 207–8

black holes 68, 69, 108, 109, 125–6, 186, 211; Black Hole Era 126; singularity 67; supermassive 126
Black Swan events 209
Bohr, Niels 86
Brahe, Tycho 20
brain-computer interfaces 213
Breakthrough Listen 191
Breakthrough Starshot 176–7
Brout, Robert 97
Bruno, Giordano 132–3
Burbridge, Margaret 113

Caltech (California Institute of Technology) 89–91, 110, 111
Cambrian explosion 144, 163–4
Carnegie Institute 113
Carnegie telescope, Palomar Observatory 113
Cassini (space probe) 166–71, 219
Catholic Church 18, 24, 25, 133
cephalopods 149–51
CERN (European Organization for Nuclear Research) 90–91, 98–100, 104
chimpanzees 145
Clarke, Arthur C. 184, 186
climate change 147, 183–4, 204, 206, 208–9
cloud chamber 89–90
Cocconi, Giuseppe 187–8
Coma Berenices 110–11
comets 4, 31–4, 141, 204, 205, 219
consciousness 55, 195, 200–201, 207, 209, 213, 220
Copernican Revolution/Copernicus, Nicolas: *On the Revolutions of the Celestial Spheres* 17–26, 27, 46
cosmic microwave background radiation 118–21, 122
cosmos, etymology of 7
CRISPR-Cas9 212
Curtis, Heber 37–8
Cygnus (constellation) 173

Dalton, John 76
dark biosphere 172
dark energy 110, 122–4, 126–7
dark matter 101–2, 109–18; dark sector and 101, 117; future of world predictions and 124–7; gravitational lensing and 115–16; Modified Newtonian Dynamics (MOND) and 118; Rubin and 112–15, 116–17; Weakly Interacting Massive Particles (WIMPs) and 117–18; Zwicky and 110–11, 115, 116
Darrigol, Olivier 53
Deep Space Network 164
Degenerate Era 125–6
Democritus 73, 132
Denisovans (*Homo floresiensis*) 147
Dicke, Robert 120, 121
Didymos (asteroid) 205
Dimorphos (asteroid) 205
dinosaurs 142, 144, 149, 204
Dione (moon) 169
Dirac, Paul 89, 90
disorder 74, 135–7
DNA 26, 138, 140, 145, 147, 194, 196, 212, 218
Double Asteroid Redirection Test (DART) 205
Drake, Frank 62
Dyson, Freeman/Dyson sphere 185, 188, 193, 203
Dysonian SETI 193

Earth (planet): antimatter on 104; Aristotelian geo-centrism (Earth as centre of universe) 8–9, 10, 15, 17–20, 23, 25, 32, 132; asteroid impacts 144–5, 204–7; Earthrise photograph (1968) 162–3; exceptionalism and 131–2; future of 199–214; heliocentric model/Copernican Revolution 17–26, 27, 46, 132; *Homo sapiens* attunement to ix, 216–20; life, evolution of on 131–52, 182–4, 200, 203–4; muons rain down on 96; Overview Effect 162–3; Pale Blue Dot photograph (1990) 162; Rare Earth hypothesis 183–4; rotation 8, 18, 46–7, 54, 59, 217; spherical shape 8–9
eclipses 4–6, 8–9, 14, 49–50, 61–3
Eddington, Sir Arthur 62–3, 127, 135
Ehman, Jerry 190
Ehrenfest, Paul 61
Einstein, Albert: education 47–9; Einstein ring 116; general relativity, theory of 48, 57–61, 62, 63, 64, 67–70, 115; 'God does not play at dice'

INDEX

comment 86; Newtonian physics, implications of theories for 70–71; Nobel Prize (1921) 81; quantum mechanics and 80–83, 86; special theory of relativity 47, 48, 49–55, 70, 80
Einstein, Hermann 48
electromagnetism 93, 97, 107
electronics 87, 164
electrons 65, 66, 78–81, 83, 85, 87, 89–91, 96, 105, 121, 189
electroweak force 93–4
elements: classical 9, 74–5; periodic table *see* periodic table
elliptical orbits 21, 30
Enceladus (moon) 166, 168–9, 171
England, Jeremy 137
Englert, François 97, 100–101
entropy 135
Epicureans 132
epicycles 10, 14, 19, 21
eukaryotes 142–3
Europa (moon) 22, 171
European Space Agency (ESA) 158; Planck mission 122
Everett, Hugh 85
exoplanets (planets outside our solar system) 172–8, 183, 191, 194
extraterrestrial life 164, 174–5, 179–96; civilisations thousands/millions of years ahead of us, imagining 184–7; Dark Forest theory 183; definitions of life and 139–41; Fermi paradox 180–84; Great Filter 181–2, 209; Great Silence 180–84; Kardashev scale 184–7, 193; METI (Messaging Extra-Terrestrial Intelligence) 193–6; Rare Earth hypothesis 183–4; Search for Extra-Terrestrial Intelligence (SETI) 187–93; self-replicating probes 180–81; terraforming 180; transcension hypothesis 183; Zoo hypothesis 183, 195

Fermi, Enrico/Fermi paradox 180, 181
Feynman, Richard: *Lectures on Physics* 108
First World War (1914–15) 62–3, 67, 208
flat rotation curve 114–15
force carriers 97
Ford, Kent 114

Frail, Dale 172
Freundlich, Erwin 62
future 197–220; cosmic perspective and 201–4; earth, attunement to and 216–20; existential risks 204–10; intelligence and 210–14

Gagarin, Yuri 153
galaxies: antimatter galaxies 103; black holes at centre of 67; dark matter and 110–11, 114–18, 122–3; fate of 124–7; future of 124–7, 202–3; gravitational lensing and 115–16; habitable zone and 175, 176, 183, 191; Local Group 202; number of 179; size of 37–8; velocity map 114–15. *See also individual galaxy name*
Galilei, Galileo: *Dialogue Concerning the Two Chief World Systems* 25, 46; Galilean moons 22–3; Galilean relativity 46–7, 49; heliocentrism and 22–5; Neptune, glimpses 36; *The Sidereal Messenger* 24; trial by Roman Inquisition (1633) 25; Universality of Free Fall 58
Galileo (spacecraft) 219
Galle, Johann Gottfried 35–6
Gamow, George 120
Geiger, Hans 78
Gell-Mann, Murray 90–91
geometry 13–14, 20–21, 32, 48
Glashow, Sheldon 93–4, 95
Gliese 445 163
gluons 92–3
Goethe, Johann Wolfgang von 21
gold foil experiment 78, 91
gravity: black holes and 109; dark energy and 110–11, 117–18, 123–4, 126–7; Einstein and 57–67, 70; four fundamental forces and 93, 97; gravitational lensing 115–16; gravitational waves 68–70; multi-dimensional universe and 107; Newton and 29–32, 43, 58–9, 60, 70
Great Debate (1920) 37–8
Great Plague (1665) 28
Great Pyramid, Egypt 96–7
Great Silence 180–84, 193
Greece, ancient 6, 7–11, 12, 19, 73, 74, 131–2
Grossman, Marcel 48

241

Halley, Edmund 32–3
Harvard computers 36–7
Hawking, Stephen 194; Hawking radiation 126
Heisenberg, Werner 86; *The Physicist's Conception of Nature* 85; Uncertainty Principle 81
Hercules 194
Herman, Robert 120
Herschel, Caroline 33–4, 113
Herschel, John 34
Herschel, William 33–5
Higgs boson 97–8, 100–101
Higgs field 97–8, 101
Higgs, Peter 97, 100–101
hominin species 145–6
Homo (genus) 145–6
Homo erectus 146
Homo heidelbergensis 146
Homo sapiens 53; earth, attunement to ix, 216–20; future of 214, 215–20; lifespans, extending 212; meaning, search for 221–3; name 214; origins and evolution of 142, 146–9, 216–19; philosophical species 222–3
Hooker Telescope 38, 110
House of Wisdom (Bayt al-Hikmah), Baghdad 12
Hubble, Edwin 38–9, 110
Hubble Space Telescope 174, 205
Hulse, Russell 68
Huygens (probe) 166–7
hydroxyl 189

IKAROS mission 177
incentive trap 177–8
Index of Prohibited Books 24
information, life as 138–9
intelligence: artificial 134, 180, 210–14, 218–20; extra-terrestrial 85, 150, 163, 164, 181–5, 187–96; evolution of 182; explosion of 211; future of 200–201, 210–14, 220; superintelligence 184, 211–12; terrestrial alien 149–52
International Astronomical Union (IAU) 113
International Centre for Theoretical Physics (ICTP) 94
International Space Station 103
Io (moon) 22, 49–50, 161

Islamic science 12–15, 94, 95
island universes 38, 125

James Webb Space Telescope (JWST) 174
John Paul II, Pope 25
Juno (spacecraft) 219
Jupiter 159–61, 162, 166, 175, 219; Great Red Spot 159, 160–61; moons 22–3, 49–50, 51, 161–2, 171; *Voyager* missions and 159–62, 165

Kardashev, Nikolai/Kardashev scale 184–6, 193
Kármán line 153–4
Kelvin, Lord 44, 120; 'Two Clouds' speech 44–5, 51
Kepler-16b 175
Kepler-453b 175
Kepler, Johannes 14, 20–21, 114
Kepler Space Telescope 173, 179

Lagrange, Joseph 76
Large Hadron Collider 98–100, 107–108
Laser Interferometer Gravitational-wave Observatory (LIGO) 69
LaserSETI 192
Last Universal Common Ancestor (LUCA) 142
Lavoisier, Antoine 75–6
Le Verrier, Urbain 35, 36, 60
Leavitt, Henrietta Swan 36–7
Leibniz, Gottfried Wilhelm 45
length contraction 53–4
Leucippus 73, 132
life: definition of 133–41; Earth, attunement to and ix, 216–20; Earth, evolution of life on ix, 141–9, 216–20; end of 201–2; extraterrestrial *see* extraterrestrial life; *Homo sapiens* origins and evolution 141–9; information, life as 138–9; lifespans, extending 212; order/disorder and 135–8; otherworldly 149–52
light: Big Bang and 121; general theory of relativity and 58–9, 61–3, 68, 115–16; gravitational lensing and 115–16; 'light quanta', Einstein idea of 80–81; propagation of 50–51; special theory of relativity and 49–54; speed of 50–54, 58–9, 68, 79, 177, 178, 180, 181, 188

INDEX

Linquist, Stefan 151
Little Dipper constellation 163
Local Group 202
Lorentz, Hendrik 53
Lowell, Percival 155
Lucretius: *On the Nature of Things* 132
Lutyen's Star 194

Magee, Bryan: *Confessions of a Philosopher* 70
magic frequencies 190
mammals, age of 144, 204
Mangalyaan 158
Mangloor, Pakistan ix
many worlds interpretation 85–6
Maric, Mileva 48
Mariner 4 mission 156
Mars (planet) 154–9, 182, 219
Marsden, Ernest 78
Martian Moons eXploration (MMX) 158–9
mass: black holes 69; dark matter and 'missing mass' 111, 115, 116, 118; Degenerate Era and 125–6; gravity and 29, 57; Higgs boson and 97, 101; 'law of conservation of mass' 75–6; quark 92; supernova explosion and 66–7
mass extinction 144–5, 204–5
mathematics: antimatter and 89; abstract language of 55, 105; Copernican Revolution and 19–21; intelligence, future of and 210; Islamic 13, 14; nature written in language of 195; Newton and 30, 31, 32; Ptolemy and 10
matter, laws of 73–87; atomic model 73–80; four-element model 74–5; nature of reality and 83–7; Periodic Table of Elements 76–8; quantum mechanics 80–83; structure of matter 74
Maxwell, James Clerk 50–51
meaning, search for ix, 3, 221, 222, 223
Mendeleev, Dmitri 76–7
Mercury 60–61, 70, 71, 201, 219
Mesopotamia 4, 154
METI (Messaging Extra-Terrestrial Intelligence) 193–4
Michelson, Albert A. 43–4
microbes 171, 207–8

Milky Way (galaxy) ix, 179, 180; age of 180; Andromeda, predicted to collide with 202; black hole at centre of 67; cosmic microwave background radiation 118–19; exoplanets in 175; Great Debate (1920) and 37–8; last stars burn out in 125, 203; Local Group and 202; SETI searches and 191; as a singular entity 37–9, 110; size of 37, 181
mind uploading 213–14
Minkowski, Hermann 53
mitochondria 143
Modified Newtonian Dynamics (MOND) 118
Moon 9, 10, 11, 17, 23, 29, 58, 62, 112, 156, 163, 174, 219; Apollo 11 landing 153–4; Apollo 15 moonwalk 58; lunar eclipse 4, 6, 8
Morrison, Philip 187–8
motion: motion of inclination 18; plantary 5–6, 8, 10, 14, 17–21, 29, 30, 32, 35, 43, 46–55, 59–61, 70–71, 114, 183, 217; prograde 10; relative 46–55; retrograde 10, 18; triple 20
Mount Wilson Observatory, California 38, 110, 113
multicellularity 143–4
multi-dimensional world 105–8
muon 96–7

nanocrafts attached to light sails 177
NASA 58, 134, 156, 157, 158, 159–60, 205
natural selection 134, 144, 148
Neanderthals 146, 147
Neddermeyer, Seth 96
Neptune (planet) 35–6, 60, 71, 159, 160–62, 219
neural networks 213
New General Catalogue (NGC) 34
Newton, Isaac 27–32; birth and childhood 27–8; Einstein's theories and end of Newtonian physics 70–72; *Principia* 30–31; Trinity College Cambridge 28; universal gravitation, laws of 29–33, 35, 36, 43, 44, 45, 58–60, 63, 64, 118; 'year of wonders' 28–9
Nobel Prize in Physics: (1921) 81; (1968) 96; (1979) 93–5; (2013) 100–101

nuclear physics/nuclear weapons 64, 65, 78, 102, 104, 119, 120, 125, 157, 206, 207
nucleus 66, 69, 79

octopus 149–51
Oliver, Bernard 189–90
order, drive to seek amid disorder 74, 135–7
oxygen 66, 75, 139, 143, 158, 174, 189

Pale Blue Dot (photograph) 162
particle accelerators 90–91, 104, 118
particle colliders 92, 96, 98–101, 102, 105, 107, 199
particle physics *see* physics
particle zoo 90, 92, 96
Pascal, Blaise: *Pensées* 32
pathogens, designer 208
Penzias, Arno 118–21
Periodic Table of Elements 66, 77, 90, 95
Perseverance (Mars rover) 157–8
photons 81, 189
photosynthesis 136
physics: classical/Newtonian 29–33, 35, 36, 43, 44, 45, 58–60, 63, 64, 70–72, 118; Nobel Prize for *see* Nobel Prize; nuclear physics 64, 65, 78, 102, 104, 119, 120, 125, 157, 206, 207; Standard Model of particle physics 95–101
Pioneer 1 (probe) 165–6, 194
Pioneer 2 (probe) 194
Planck, Max 81, 122
planets: Aristotelian cosmological system *see* Aristotle; Copernican cosmological system *see* Copernican Revolution/Copernicus, Nicolas; discovery of new *see individual planet name*; exoplanets *see* exoplanets; extraterrestrial life and *see* extraterrestrial life; Kepler's laws of planetary motion 20–21, 30, 114; motion of 5–6, 8, 10, 14, 17–21, 29, 30, 32, 35, 43, 46–55, 59–61, 70–71, 114, 183, 217; name/term 8; Ptolemy and motion of 10, 14; space exploration and *see* space exploration; spacetime and 59–60; transit of, Babylonians track 5–6
Plato 8

'plum pudding' model 78–80
plurality of worlds 131–3
Poincaré, Henri 53
Popper, Karl 71
Porco, Carolyn 162
positron 89–90, 104–5; positron emission tomography (PET) 104–5
Priestley, Joseph 75
Princeton University 120
prograde motion 10
programmability 11
Project Ozma 187
protected spectrum 189
Protestant Reformation 24
Proxima Centauri 176–7
Ptolemy: *Almagest* 10, 14, 18
pulsar 68, 164, 172

quantum mechanics 44, 45, 80–83, 85–7, 89, 92, 93, 122, 189, 195, 199; computing, quantum 87; Copenhagen interpretation 83, 85; dark energy and 124; entanglement 81–2; measurement problem 83, 85; nature of reality and 83–7; proton and 92; quantum probabilities 82–3; Schrödinger and 135, 136; success of 87; Uncertainty Principle, Heisenberg's 81
quarks 91–3, 96

Rabi, I. I. 96
radio astronomy 68, 118–20, 172, 179, 187–92
radio waves 187–8, 192–3
reality, nature of x, 83–7
relativity 45–8; death of stars and 64–7; end of Newtonian physics and 70–71; Galilean 46–7, 49; general theory of 48, 57–61, 62, 63, 64, 67–70, 115; gravitational waves and 68–9; motion, relative 46–55; special theory of 47, 48, 49–55, 70, 80
Renaissance 10, 17, 18
robotics 212
Roemer, Ole Christensen 49–50
Ross 248 (star) 163
Royal Astronomical Society 34–5, 113
Royal Institution 44, 78
Royal Observatory, Paris 49–50
Royal Society 27, 33

Taylor, Joseph 68
technology: biological 208; future of 87, 177–8, 180, 184–6, 200–201, 208, 210–14, 220; leaps in 99–100, 146, 157, 164, 177, 180, 182, 184, 211; singularity, technological 211; technosignatures 187–8, 192–3, 194
Tereshkova, Valentina 153
terraforming 180
TESS (Transiting Exoplanet Survey Satellite) 173–4
translation movement (Harakat al-Tarjama) 12
theory of everything 105
thermodynamics 44, 120; second law of 135–8
35P/Herschel-Rigollet (comet) 34
Thomson, Joseph John (J. J.) 78
Tianwen-1 (Mars mission) 158
time 9, 43, 45; base-60 notation 5–6; 'flow of time' 55; Newton and 45; spacetime 44, 52–5, 57–62, 67–70, 115–16, 124–5, 186–7; time dilation 53–5; timekeeping, cosmos used as tool for 3, 5–6, 12, 13
Titan (moon) 160, 161, 166–8
Toba volcano 206
transistors 87
transit method 173
TRAPPIST–1 system 191
trigonometry 13
Turner, Matthew 33
Tusi couple 14

unifying theory of nature 93–5
universe: age of 122; Big Bang/origins of 98, 99, 101–2, 120–21, 126, 176, 211; Big Freeze/Heat Death of 126; Big Rip 127; dark matter and *see* dark matter; end of 124–7, 203; expanding 39, 119–26, 131, 203; future of 124–7, 201–5; geo-centrism, Aristotelian 8–9, 10, 15, 17–20, 23, 25, 32, 132; heliocentric model/ Copernican Revolution 17–26, 27, 30, 46, 132; island universes 38, 125; mechanistic 43; multi-dimensional 105–8; size of 37–9; unified theory of 74, 105
unmoved mover 9
Uranus (planet) 33, 35, 60, 159–62, 219

Venus (planet) 23, 162, 177, 201, 219
volcanoes 97, 156, 161, 182, 204, 206, 207, 209
von Neumann probe 180
Voyager 1 spacecraft 159–66, 178, 194, 219
Voyager 2 spacecraft 159–66, 194, 219
Vulcan (hypothetical planet) 60–61

water: classical elements and 74, 75; extraterrestrial life and presence of 140, 156–8, 160, 165–71, 176, 183
waterhole (band of frequencies) 189–90
weak nuclear force 93, 97, 102
Weinberg, Steven 93–4, 95
Wheeler, John 59
Wilson, Robert 118–21
Winteler, Jost 47–8
Wolszczan, Aleksander 172
women, astronomy and 36–7, 48, 112–15, 116
Wow! Signal 190

Yucatán Peninsula asteroid strike 144–5, 204

Zaitsev, Alexander 193
Zweig, George 90–91
Zwicky, Fritz 110–11, 115, 116

INDEX

Rubin, Robert 112–13
Rubin, Vera 112–15, 116–17
Rutherford, Ernest 78–9, 91

Sagan, Carl 162, 163
Sagittarius (constellation) 190
Salam, Abdus 93–5
Sandage, Allan 113
Saros cycle 6
Saturn (planet) 33, 35, 159, 160–61, 165–72, 219
ScanPyramids project 97
Schiaparelli, Giovanni 155
Schrödinger, Erwin: *What Is Life?* 135–6
Schwarzschild, Karl 67
science: collaboration and 98, 101; counterintuitive reality and 86; discarded theories 84; scientific method 13–14, 27; Scientific Revolution 27; spirituality and 13–14, 21, 25, 94–5
science fiction 154, 184, 210
Scott, David 58
Search for Extra-Terrestrial Intelligence (SETI) 187–93
Seleucus of Seleucia 19
self-replicating probes 180–81
sense perceptions 216–17
sexagesimal (base-60) system of arithmetic notation 5–6
Shapley, Harlow 37–8
singularity 67; technological 211
Small Magellanic Cloud 37
solar sail propulsion 177
solar system: size of 35, 172; *Voyager* 'family portrait' of 162
Solvay Conference (1927) 86
space exploration 153–78; Apollo missions 58, 153–4, 162–3; *Cassini* 166–71; Earth 2.0 173–4; exoplanets (planets outside our solar system) 172–8; *Huygens* probe 166–7; Mars and 154–9; 'ocean worlds' 165–72; overview effect 162–3; Valentina Tereshkova/first woman to breach Kármán line 153; *Voyager* spacecraft 159–66, 178, 194, 219; Yuri Gagarin/first human to breach Kármán line 153
spacetime 44, 54–5, 57, 59–62, 67–70, 115–16, 124–5, 186–7
Stachel, John 49

Standard Model of particle physics 95–8, 101
Stanford Linear Accelerator 91–2
stars: antimatter and 103; binary star systems 175–6; Cepheid variable stars 37–8; closest to Earth (Proxima Centauri) 176–7; dark matter and 114–15; death of 64–7, 68, 123, 124, 125–7, 172–3; Degenerate Era and 125–6; Earth, future of and 201–3; failed stars 125–6; fixed stars 8, 9, 18, 33; Goldilocks zone and 170; human origins and 139–40; METI and 193–4; Milky Way, number of in 179–80; nebulous 34, 38–9, 64; origins of 123, 125–6; planets tidally locked to 175–6; pulsar 68, 164, 172; rapture at sight of 220; red dwarf 125, 163, 194; red giant 65, 125, 201; SETI and 187–93; spacetime and 59–63, 70; star clusters 34, 36; starlight, bending of 61–3, 70; starlight, journey of 50; Stelliferous Era 125; velocities of in galaxies 114; white dwarf 65. *See also individual star name*
string theory 105–6
strong nuclear force 93, 97
subatomic world x, 53, 78, 87, 89–108, 117, 126, 199; antimatter 101–5; multi-dimensional universe 105–8; particle colliders 98–101; particle zoo 90–93; Standard Model of particle physics and 95–8; unifying theories and 93–5
Sun (planet): Aristotelian geo-centrism 8–9, 10, 15, 17–20, 23, 25, 32, 132; Babylonians track movement of 4, 5; death of 58–9, 65–7, 201–2; Degenerate Era and 125; heliocentric model/Copernican Revolution 17–20, 27, 30, 46, 132; Islamic science traces movement of 13; Kepler's laws of planetary motion and 21, 30, 114; Mars and 154, 155, 157; Milky Way and 37–8; nuclear fusion powering 102; origins of life on Earth and 141, 143; Ptolemy tracks movement of 10; solar eclipse 4, 61, 62; spacetime and 59–63
supernova 66, 125, 172
supervolcanism 182, 206

245